$a^2 + b^2 = c^2$

I0076899

COLLECTION OF MATHEMATICS

PROBLEMS FROM PREALGEBRA

TO CALCULUS,

WITH INSPIRATIONAL AND
MOTIVATIONAL QUOTATIONS

Irie Glajar

$sin^2x + cos^2x = 1$ $\qquad\qquad \int f'(x)dx = f(x) + C$

COLLECTION OF MATHEMATICS

PROBLEMS FROM PREALGEBRA

TO CALCULUS,

With Inspirational and Motivational Quotations

Irie Glajar

Published By
Positive Imaging, LLC
http://positive-imaging.com
bill@positive-imaging.com

All Rights Reserved

ISBN 9781944071295

Also by Irie Glajar

WE ARE ALL ONE, The End of All Worries:
Scientific and Spiritual Testimonies to the Unity of All Things
http://the-end-of-all-worries.com

TEACH FOR LIFE,
Essays on Modern Education for Teachers, Students, and Parents
http://teachforlife.positive-imaging.com

ESCAPE TO FREEDOM,
Chronicles of a Life on Two Continents,
My Escape from Communist Romania,
an Autobiography
http://escapetofreedom.positive-imaging.com

2015 INSPIRATIONAL MATHEMATICS CALENDAR AND DAY PLANNER
http://mathcalendar.positive-imaging.com

EDUCATION IN A CHANGING WORLD,
Essays for a Better Life"
http://educationinachangingworld.positive-imaging.com

All are also available on Amazon.com

Table of Contents

Acknowledgements

Looking back over more than five decades of my learning and teaching mathematics, I want to express my highest gratitude to some of the esteemed people involved.

I owe my initial formation to my parents, Doina and Ioan Glajar, and my Ucea de Jos, Romania, elementary school teacher from first to fourth grade, Mrs. Olimpia Barbat. My parents' insistence on success, combined with my teacher's professional care helped instill in me gradually a deep love for mathematics.

Eventually, with the clear explanations and unconditional dedication of my high school of Victoria math teacher, Mr. Laurentiu Comsoiu, I was able to successfully propel myself toward higher education in mathematics within a very competitive academic environment of the 1974 Romania.

At the University Babes-Bolyai of Cluj-Napoca, a host of specialized professors, some internationally renowned academicians, have shaped my higher understanding of mathematics. In this respect, Dr. Andrei Ney, Dr. Francisc Rado, and Dr. Dimitrie D. Stancu have been especially inspirational through their wider and more practical approach to the teaching of mathematics at the graduate level.

Since my arrival in Austin, Texas in 1982, when I started teaching mathematics at two private high schools, continuing at the Austin Community College and the University of Texas at Austin (for many years at two institutions simultaneously), my students have earned all my gratitude. They have been challenging me to constantly adjust my explanations to their understanding, and, even today, they reinforce my irrevocable conviction that a humane approach to education should be embraced at all costs.

As my present collection of 1,460 math problems is concerned, I want to extend sincere gratitude to Mr. Tom Beversdorf who with unbelievable dedication meticulously edited the entire set of problems. Any still possible discrepancies are my responsibility, and I would appreciate reader's feedback in order to rectify them. For comments please use my email: ir_gl@yahoo.com, and the subject line "Math problems."

Introduction

The investigation of mathematical truths accustoms the mind to method and correctness in reasoning, and is an employment peculiarly worthy of rational beings.

George Washington

Ever since 1984, when I started my teaching career in mathematics at Austin Community College, Austin, Texas, I envisioned writing a collection of math problems meant to cover most of the undergraduate spectrum. That aim led me to publish my "2015 Mathematics Calendar" which consisted of four math problems per day, accompanied by a suggestive inspirational quote, and with the answers in the respective date (example: 1-1-20-15).

In order to make this collection of math problems available to those interested for generations to come, I have decided to rearrange it into the present format. Therefore, the 1,460 problems are presented in groups of four, numbered such that problems 1 and 2 cover developmental mathematics (the essential elementary and secondary school curriculum), while numbers 3 and 4 cover most of the college credit sequence from college algebra to calculus. Each group of four starts with a meaningful quotation and ends with the answers to the *last question* in each of the four problems in the order 1, 2, 3, 4, like this: (4-26-20-15). So, all answers are natural numbers.

In addition, this volume starts with my proofs of the Pythagorean Theorem and some of the basic divisibility rules, important mathematical concepts that are usually presented to students without pertinent justification. In this respect I focused on divisibility by 3, 9, and 4 for which my proofs follow virtually the same pattern. I hope the readers will find them interesting and helpful.

In the end, as we all know, practice doesn't make it perfect but it makes it better. This book can help those dedicated to their improvement in the realm of undergraduate mathematics simply by practicing consistently, patiently, joyfully, and persistently. How to use this collection of math problems is left to each individual's decision according to her/his readiness, pleasure, and need. Moreover, the 365 inspirational/motivational quotations included here are meant to help one combine mathematical and philosophical thinking, which can definitely contribute to significant personal advancement on both fronts.

Irie Glajar
Mathematics Professor
Austin Community College, Austin, Texas
irieg@austincc.edu or ir_gl@yahoo.com.

Proofs by Irie Glajar

A Proof of the Pythagorean Theorem

Let ABC be a right triangle with B being the right angle. First, we drop the height BD with point D on AC. From here on, for the sake of simplicity, we will use these notations:

$$AB = a, BC = b, AC = c, BD = h, AD = t, \text{ and } DC = c - t. \quad (0)$$

To prove the Pythagorean Theorem we will use the area of a triangle formula (area = half of base times height) and similarity of triangles. Along the way, as a byproduct of this proof, I will point out three other useful relationships in a right triangle.

Let's begin by expressing h in terms of the sides of the triangle, a, b, c. For this, we use the fact that the area, S, of triangle ABC can be expressed in two different ways: $S = \frac{ab}{2}$ or $S = \frac{ch}{2}$. Now we set $\frac{ab}{2} = \frac{ch}{2}$ that is equivalent to ab = ch, in which we can easily solve for h:

$$h = \frac{ab}{c} \quad (1)$$

(It is obvious that (1) is a useful relationship in a right triangle.)

Since triangle ABD is similar to triangle BCD, their sides are proportional:

$$\frac{BD}{DC} = \frac{AD}{BD} = \frac{AB}{BC}$$

Using our notations from (0) and the relationship in (1), we can write these proportions this way:

$$\frac{\frac{ab}{c}}{c-t} = \frac{t}{\frac{ab}{c}} = \frac{a}{b} \quad (2)$$

Solving for t in the second proportion from (2), we obtain: $t = \frac{a^2}{c}$ (3)

Solving for (c-t) in the first proportion from (2), we get: $c-t = \frac{b^2}{c}$ (3')

(We can see that (3) and (3') are two other useful relationships in a right triangle.)

Now we substitute (3) into the first proportion from (2), and we have:

$$\frac{\frac{ab}{c}}{c-\frac{a^2}{c}} = \frac{\frac{a^2}{c}}{\frac{ab}{c}} \qquad (4)$$

By cross-multiplying and simplifying in (4), we get:

$$\frac{(a^2b^2)}{c^2} = a^2 - \frac{a^4}{c^2} \qquad (5)$$

In order to clear the fractions, we now multiply (5) by c^2, and we have:

$$a^2b^2 = a^2c^2 - a^4 \qquad (6)$$

Now we divide (6) by a^2, and we get:

$$b^2 = c^2 - a^2 \qquad (7)$$

Finally, by adding a^2 to both sides of (7), we can arrange (7) in the traditional form of the famous Pythagorean Theorem in a right triangle with legs a and b, and the hypotenuse c: $a^2 + b^2 = c^2$.

Proof of Divisibility by 3

Part 1

Let's start with an example of the divisibility by 3 rule: 246 is divisible by 3 if the sum of its digits is divisible by 3; because $2 + 4 + 6 = 12$, and 12 is divisible by 3, we know that 246 is divisible by 3, or that 246 is a multiple of 3; indeed, $246 = 3 \cdot 82$. The question is: why does this rule work every time? In the following we will provide an answer to this question.

Part 2

Let's verify this rule for an integer number of three digits in general; then, we will prove that the divisibility by 3 rule works for any integer. For the sake of simplicity, let's write this number like this:

$N = a_1a_2a_3$, where a_1, a_2, and a_3 are the three digits of the number.

Therefore, we can write N in expanded form this way:
$$N = 100a_1 + 10a_2 + a_3. \qquad (0)$$

The condition we want to verify is that if $a_1 + a_2 + a_3 = 3k$, where $k = 1$, 2, 3, ..., then N is divisible by 3 (or N is a multiple of 3).

To accomplish this, we isolate $a_3 = 3k - a_1 - a_2$ and we substitute it in (0):
$$N = 100a_1 + 10a_2 + 3k - a_1 - a_2$$

Combing like terms, in simplest form N is: $\qquad N = 99a_1 + 9a_2 + 3k$.
Factoring out 3 as a GCF of the right side, we obtain: $N = 3(33a_1 + 3a_2 + k)$, which shows that N is a multiple of 3 since $(33a_1 + 3a_2 + k)$ is an integer.

Part 3

Let's now prove that *any integer of n digits whose digits add up to a multiple of 3 is itself a multiple of 3*. Let $N = a_1a_2a_3...a_n$.

Then, the divisibility by 3 condition (the sum of the digits is a multiple of 3) can be written like this:

$\sum_{i=1}^{n} a_i = 3k$, where k = 1, 2, 3, ... (1)

In (1) we solve for the last digit:

$a_n = 3k - \sum_{i=1}^{n-1} a_i$. (2)

Now we write our number N in expanded form according to the place values of its digits and using (2) as a substitute for the last digit:

N = $\sum_{i=1}^{n-1}(10^{n-i} a_i) + 3k - \sum_{i=1}^{n-1} a_i$. (3)

Since the two summations in (3) can be consolidated into one, and by factoring out every a_i, we obtain:

N = $\sum_{i=1}^{n-1}[(10^{n-i} - 1)a_i] + 3k$. (4)

At this point it's obvious that $(10^{n-i} - 1)$ in (4) are all multiples of 9 for any n-i ≥ 1, which, by using summation to express the bracket in (4), allows us to write N like this:

N = $\sum_{i=1}^{n-1}[(\sum_{j=0}^{n-i-1}(9 \cdot 10^j))a_i] + 3k$. (5)

Properties of the summation notation allow us now to factor 3 as a GCF and write it in front, like this:

N = 3· $\sum_{i=1}^{n-1}[(\sum_{j=0}^{n-i-1}(3 \cdot 10^j))a_i] + 3k$. (6)

Finally, we can factor out 3 from both terms of the right side of (6), which, since the simplified value of the expression in the braces is an integer, it proves that N is a multiple of 3:

N = 3· $\{\sum_{i=1}^{n-1}[(\sum_{j=0}^{n-i-1}(3 \cdot 10^j))a_i] + k\}$.

Proof of Divisibility by 9

Since the rule for divisibility by 9 is very similar to that of divisibility by 3, we will continue with the previously established relations: *if the sum of the digits of the given number N is divisible by 9, then N is also divisible by 9*. To prove this, the only difference is that, in the proof for divisibility by 3, from the start we set up (1), the sum of the digits of the number N, like this:

$$\sum_{i=1}^{n} a_i = 9k, \text{ where k = 1, 2, 3, ...} \tag{1'}$$

Therefore, (5), using (1'), becomes:

$$N = \sum_{i=1}^{n-1}\left[\left(\sum_{j=0}^{n-i-1}(9 \cdot 10^j)\right)a_i\right] + 9k. \tag{7}$$

It is now obvious that in (7) we can factor out 9, and obtain:

$$N = 9 \cdot \left\{\sum_{i=1}^{n-1}\left[(10^j))a_i\right] + k\right\}. \tag{8}$$

This proves the rule for divisibility by 9, since the simplified value of the expression in the braces in (8) is an integer.

Proof of Divisibility by 4

Part 1

The rule for divisibility by 4 states that *if the last two digits of an integer number N of 2 or more digits form a number M that is divisible by 4, then the number N is also divisible by 4.* Here is a simple example: the number 2432 is divisible by 4 since 32 is divisible by 4 (or we can say 2432 is a multiple of 4, which in fact it's true: 2432 = 4 · 608).

Part 2

Let's verify this rule for a number of 4 digits in general: N = $a_1 a_2 a_3 a_4$. First, we write N in the expanded form according to the place values of its digits:

$$N = 1000a_1 + 100a_2 + 10a_3 + a_4 \qquad (0)$$

The rule is that if M = $a_3 a_4$ is a number divisible by 4, N is also divisible by 4. To show this, we write M in expanded form and set it as a multiple of 4 this way:

$$M = 10a_3 + a_4 = 4k, \text{ where } k = 1, 2, 3, \dots \qquad (1)$$

Then we solve for a_4 in (1), and we have a_4 = 4k - 10a_3. By substituting it in (0) we get:

$$N = 1000a_1 + 100a_2 + 10a_3 + 4k - 10a_3, \text{ which in simplest form is:}$$

$$N = 1000a_1 + 100a_2 + 4k. \qquad (2)$$

By factoring out 4 in the right side of (2) we obtain:

N = 4(250a_1 + 25a_2 + k), which, since the simplified value of the expression in the parentheses is an integer, proves that N is a multiple of 4, or that N is divisible by 4.

Part 3

We will prove now that if the last two digits of any integer N form a number M that is divisible by 4, then N is also divisible by 4 (or N is a multiple of 4).

Let N = $a_1a_2a_3...a_{n-1}a_n$.

We will prove that if M = $a_{n-1}a_n$ is divisible by 4, then N is also divisible by 4.

The condition for this divisibility can be written like this:

M = $a_{n-1}a_n$ = 4k, where k = 1, 2, 3, ... Then, in expanded form we can write it this way:

$$10a_{n-1} + a_n = 4k \tag{1}$$

In (1) we solve for a_n:

$$a_n = 4k - 10a_{n-1} \tag{2}$$

Now we write N in expanded form using the summation notation according to its place values of its digits and substituting a_n using (2):

$$N = \sum_{i=1}^{n-1}(10^{n-i} a_i) + 4k - 10a_{n-1} \tag{3}$$

Since the last term of the summation is $10a_{n-1}$, we can write (3) like this:

$$N = \sum_{i=1}^{n-2}(10^{n-i} a_i) + 10a_{n-1} + 4k - 10a_{n-1} \tag{4}$$

After combining like terms, we obtain:

$$N = \sum_{i=1}^{n-2}(10^{n-i} a_i) + 4k \tag{5}$$

In (5) we can factor 10^{n-i} as $2^{n-i}5^{n-i}$. Now, it is obvious that 2^{n-i} are all multiples of 4 for all i = 1, 2, 3, ..., n-2. Therefore, since $4 = 2^2$ we can factor out a 2^2 from the entire summation, and we can write (5) this way:

$$N = 2^2 \sum_{i=1}^{n-2}(2^{n-i-2}5^{n-i} a_i) + 4k \qquad (6)$$

Finally, since $2^2 = 4$, in (6) we can factor out a 4, which proves that N is a multiple of 4 since the simplified value of the expression in the bracket is an integer:

$$N = 4 \left[\sum_{i=1}^{n-2}(2^{n-i-2}5^{n-i} a_i) + k\right]$$

**Mathematics Problems from Prealgebra to Calculus,
with Inspirational and Motivational Quotations**

Important note: in each group of four, problems numbered 1 and 2 are developmental, while those numbered 3 and 4 are college credit problems (please see the introduction and the table of contents).

Prealgebra (1, 2) and College Algebra (3, 4)

We are what we repeatedly do.

Aristotle

1. Indicate the digit that represents hundreds in the following number: 3,098,174.

2. Specify the digit that stands for hundredths in the number: 22.013.

3. Let $f(x) = \frac{x}{60-3x}$. For what value of x is the function undefined?

4. Find the left-side limit of the interval notation for the domain of the function:
   $f(x) = \sqrt{3x - 45}$.

(1-1-20-15)

We are a way for the universe to know itself.

Carl Sagan

1. How many words are there in the "word name" of the number "5"?

2. How many digits larger than 7 are there in the number 7,349,568?

3. Find the real zero of the polynomial function
   $f(x) = x^3 - 21x^2 + 21x - 20$.

4. How many whole numbers less than 15 are there in the range of the function

   $f(x) = \sqrt{256 - x^2}$?

(1-2-20-15)

No man remains quite what he was when he recognizes himself.
Thomas Mann

1. What is the digit that represents hundreds after you round 4,149 to the nearest hundred?

2. Find the digit you need to change if you would round 2,300 to the nearest thousand.

3. Determine the value of k if the point (- 2, k) is on the graph of $h(x) = 3x^2 - 4x$.

4. Find the positive x-intercept of the function $g(x) = 2x^2 - 27x - 45$.

(1-3-20-15)

Somehow we learn who we really are and then live with that decision.
Eleanor Roosevelt

1. How many digits less than 6 but different than 0 are left in the number 7,854,379 if you would round it to the nearest ten-thousand?

2. What is the new digit different than 0 in the number 2,365,876 if you are to round it to the nearest hundred-thousand?

3. Find the y-intercept of the parabola $f(x) = (3x - 4)(2x - 5)$.

4. Find the x-intercept of the line 4x - 60 = - 5y.

(1-4-20-15)

Everyone, without exception, is searching for happiness.
Blaise Pascal

1. Use the associative property to simplify $5 + (3 + (-7))$.

2. Use the commutative property of addition to simplify
   $7 - [(-3) + 5]$.

3. What is the sum of the x- and y-coordinates of the center of the circle
   $$x^2 + y^2 - 14x - 26y + 217 = 0?$$

4. Find the radius of the circle $x^2 + y^2 = 2x - 4y + 220$.

(1-5-20-15)

Happiness is the meaning and the purpose of life, the whole aim and end of human existence.
Aristotle

1. Use the associative and/or commutative properties to simplify
   $3 - [7 + (-6 + 1)]$.

2. Simplify $[7 - (5 - 3)] + [3 + (-1 -1)]$.

3. Determine the hypotenuse of a right triangle of area 96 if one leg is 12.

4. Find the absolute value of the slope of the line passing through the points of coordinates $(-9, 60)$ and $(-4, -15)$.

(1-6-20-15)

Pleasure is the object, duty and the goal of all rational creatures.

Voltaire

1. Evaluate the expression $x + y - 15$ if $x = 9$ and $y = 7$.

2. Evaluate the expression $31 - a - b$ if $a = 13$ and $b = 11$.

3. A trapezoidal flower garden with bases 7ft and 12ft has the area $190\ ft^2$. Find the shortest distance between its bases.

4. A store is offering a 60% discount on tennis shoes. Write a linear function that gives the sale price S in terms of the list price L and find the sale price of a pair listed at $37.50.

(1-7-20-15)

We all live with the objective of being happy; our lives are all different and yet the same.

Anne Frank

1. Translate using mathematical symbols, then evaluate if x is 19: "37 less than two times x."

2. Evaluate "- 9 plus y" if y is 17.

3. Find the positive solution of the equation: $-x(18 - x) = 40$.

4. Determine the positive zero of the function
   $f(x) = 3x^4 - 45x^3 + 3x - 45$.

(1-8-20-15)

It is a common experience that a problem difficult at night is re-solved in the morning after the committee of sleep has worked on it.

John Steinbeck

1. By what value should we extend the length of a 5 by 8 rectangle for the new perimeter to be 28?

2. The perimeter of a square is 36. Find the length of a side.

3. Use the quadratic formula to solve the equation:
   $x^2 = -40(10 - x)$.

4. Solve the rational equation: $\dfrac{3}{x-2} - \dfrac{64}{x^2-4} = -\dfrac{1}{x+2}$.

(1-9-20-15)

Do not shorten the morning by getting up late; look at it as the quintessence of life, and to a certain extent sacred.

Arthur Schopenhauer

1. Simplify: $-2x + 6 + (2x - 5)$.

2. Simplify: $3(y + 6) - 3y - 8$.

3. Find the value of the function $f(x) = -x^2 + 5x + 34$ for $x = 7$.

4. For what value of x is $g(x) = -15$ if $g(x) = \dfrac{3x}{12-x}$.

(1-10-20-15)

Children are likely to live up to what you believe of them.
Lady Bird Johnson

1. Evaluate the expression 3x5 − 11y4 for x = 3 and y = 1.

2. Evaluate the expression 3xy − 3x − 4y for x = 3 and y = 4.

3. For what real number is the equation undefined:

$$\frac{x}{4} = \frac{x+2}{3x-60}.$$

4. Determine the sum of the solutions of the equation:
$$2x^3 = x(30x − 72).$$

(1-11-20-15)

Trust men and they will be true to you; treat them greatly and they will show themselves great.
Ralph Waldo Emerson

1 Simplify (- 1)(1)(- 1)(1)(- 1)(- 1).

2 If the sides of a triangle are x + 1, x + 2, and x + 3, find its perimeter for x = 2.

3 Leo works twice as fast as Mack. If working together they take 6 hours and 40 minutes to finish a job, how long would it take Mack to do the same job alone?

4 A quadratic function has x = - 6 as an x-intercept. Find the other x-intercept if the points of coordinates (14, - 20) and (16, 22) are on the graph of the function.

(1-12-20-15)

Any device whatever by which one frees himself from fear is a natural good.

Epicurus

1   Simplify: $\frac{17-9}{32-24}$.

2. Simplify: $\frac{126-9}{9}$.

3. Use the quadratic formula to solve the equation and list only the positive solution: $2(x^2 - 10) = 39x$.

4. Find the smaller of the two solutions of the equation: $x(31 - x) = 240$.

(1-13-20-15)

To persevere, trusting in what hopes he has, is courage. The coward despairs.

Euripides

1. Simplify: $\frac{13+22}{7 \cdot 5}$.

2. Simplify: $\frac{37-9}{6-4}$.

3. Solve: $16 + \sqrt{x - 4} = x$.

4. Find the positive solution of the equation:
    $$\sqrt{2x + 6} - 1 = \frac{x}{3}.$$

(1-14-20-15)

Choose always the way that seems the best, however rough it may be; custom will soon render it easy and agreeable.

Pythagoras

1. Simplify: $5^2 - 6(4)$.

2. Simplify: $2 + 3(7 - 3) + 1$.

3. Determine the larger of the solutions of the absolute value equation:

$$|2x - 25| - 15 = 0.$$

4. Find the integer solution of the equation: $|3x + 2| = |4x - 13|$.

(1-15-20-15)

If, after all, men cannot always make history have a meaning, they can always act so that their own lives have one.

Albert Camus

1. Simplify: $6(7) - 6^2 - 5$.

2. Simplify: $1 + 3^2 + 2(3) - 12 + 4(3)$.

3. Find the absolute value of the product of the real solutions of the absolute value equation: $|x + 20| = x^2$.

4. Find the positive rational solution of the equation:
$x^2 - |2x + 9| = 186$.

(1-16-20-15)

We, ignorant of ourselves, beg often our own harms, which the wise powers deny us for our good.

William Shakespeare

1. Simplify: $3 + 5(3^2 - 8) - 7$.

2. Simplify: $6^2 - (3^2 - 4) - 2^2 - 2(5)$.

3. Determine the right-hand-side limit of the solution for the absolute value inequality: $|2x - 25| \leq 15$.

4. Find the smallest integer larger than 12 that satisfies the absolute value inequality: $|3x - 40| > 4$.

(1-17-20-15)

The man who does not read has no advantage over the man who cannot read.

Mark Twain

1. Evaluate the expression for x = 3: $\frac{x^2 - 5}{x+1}$.

2. Evaluate the expression for x = 2 and y = 5: $\frac{5x + y^2 + 1}{y - x - 1}$.

3. Find the number of integers different than 0 in the domain of the expression:

$$\sqrt{200 - 2x^2}.$$

4. Determine the right-side limit of the domain of: $\sqrt{60 - 4x}$.

(1-18-20-15)

If you don't have time to read, you don't have the time (or the tools) to write. Simple as that.

Stephen King

1. Evaluate the expression for x = 3, y = 1, and z = 4:
$$\frac{3x-4y+z}{x+y+z+1}.$$

2. Evaluate the expression for x = 5: $\frac{x^2-x-1}{x-4}$.

3. What is the positive value of x if the distance between the points of coordinates (x, - 2) and (5, 1) is $3\sqrt{26}$?

4. Find the sum of the coordinates of the midpoint of the segment given by its endpoints (16, 6) and (- 2, 10).

(1-19-20-15)

A meaning of life: joy, reached not at the expense of others.

Irie Glajar

1. The perimeter of a rectangle is 28. If the length is 13, what is the width of the rectangle?

2. The length of a rectangle is 2x + 3y and its width is x + y. Find the perimeter if   x = 2 and y = 1.

3. Find the y-intercept of a line that has the slope -3 and passes through the point of coordinates (7, - 1).

4. A line has its y-intercept -10 and it is perpendicular to the line $y = -\frac{3}{2}x + 7$. Find its x-intercept.

(1-20-20-15)

There's one blessing only, the source and cornerstone of beatitude: confidence in self.

Marcus Annaeus Seneca

1. Four times the perimeter of a square is 16. Find the length of its side.

2. The sides of a triangle are x + 3, x + 5, and x + 1. Find its perimeter for x = 4.

3. Find the product of the x- and the y-coordinates of the common point of the lines of equations 2x + y = 13 and y = x + 1.

4. The perimeter of a rectangular flower garden is 46ft and its area is 120ft². Find the length of the longer side of the garden.

(1-21-20-15)

The greater the obstacle, the more glory in overcoming it.

Moliere

1. Solve the equation: 2 + (7 + x) = 10.

2. Solve the equation: (3y + 7) – 2y = 29.

3. Determine the y-intercept of the line passing through (-10, 15), parallel to the line of equation -3x + 6y – 2 = 0.

4. Find the value of y such that the slope of the line passing through the points of coordinates (-4, y) and (-2, 7) is -4.

(1-22-20-15)

We create our fate every day... most of the ills we suffer from are directly traceable to our own behavior.

Henry Miller

1. Solve the equation: $3x + x + 2x + 5 = 11$.

2. Solve the equation: $4y + (3y - 4) - 6y = 19$.

3. The profit of a company in the second quarter was $21,600. By what percentage did the profit increase if in the first quarter the profit was $18,000?

4. For what value of x is y = 4, if $y = \sqrt[3]{x + 49}$.

(1-23-20-15)

At thirty a man should know himself like the palm of his hand, know the exact number of defects and qualities .... And, above all, accept these things.

Albert Camus

1. Solve the equation: $3x - (x + 2) - 5 = 5 - 10$.

2. First job pays $20 per hour less than twice of what a second job pays. If the first job pays $28 per hour, how much does the second job pay?

3. Find the value of x such that the point of coordinates (x, 6) is on the graph of the function: $g(x) = 2\sqrt[3]{3x - 33}$.

4. Determine the y-intercept of the function: $f(x) = -3\sqrt[3]{5(x - 25)}$.

(1-24-20-15)

Life belongs to the living, and he who lives must be prepared for changes.

Johann von Goethe

1. Evaluate: $- (- (- (3 - 4)))$.

2. Evaluate $- (- (- (- x + 24)))$ for $x = 49$.

3. Find the absolute value of the product of the x- and y-coordinates of the center of the circle: $(x - 4)^2 + (y + 5)^2 = 49$.

4. Determine the difference between the length of the radius of the circle and the y-coordinate of the center:
$(x + 1)^2 + (y - 1)^2 = 256$.

(1-25-20-15)

Many a man curses the rain that falls upon his head, and knows not that it brings abundance to drive away hunger.

Saint Basil

1. Evaluate: $a - b - 3$ for $a = -3$ and $b = -7$.

2. Evaluate: $17 - x - y$ for $x = -10$ and $y = 1$.

3. Find the radius of the circle: $x^2 + y^2 - 6x + 4y - 387 = 0$.

4. What is the difference between the x- and the y-coordinates of the center of the circle of equation $x^2 + y^2 - 30x + 161 = 0$?

(1-26-20-15)

Misfortunes occur only when a man is false … Events, circumstances, etc., have their origins in ourselves. They spring from seeds which we have sown.

Henry David Thoreau

1. Find the value of the expression $(3x - 4y)$ for $x = -5$ and $y = -4$.

2. Evaluate $3(x - 2) + 2(y + 1)$ for $x = 9$ and $y = 2$.

3. A new line passing through the point of coordinates $(6, 11)$ is parallel to the line of equation $3x + 2y = 5$. Find the y-intercept of the new line.

4. Determine the left-side limit of the domain of the function: $f(x) = -\sqrt{\dfrac{x^2 - 13X - 30}{x+2}}$.

(1-27-20-15)

Emulation admires and strives to imitate great actions; envy is only moved to malice.

Honore de Balzac

1. Simplify: $\dfrac{(-3)(-2)(-4)}{(12)(-2)}$.

2. Find the difference in temperature from -18 degrees to 10 degrees..

3. Determine the value of x if the point of coordinates $(x, 4)$ is on the graph of the function $F(x) = \sqrt[3]{\dfrac{4x-16}{x-19}}$.

4. Find the value of y if the point of coordinates $(5, y)$ is on the graph of the parabola $y = -x^2 + 10x - 10$.

(1-28-20-15)

The greatest use of life is to spend it for something that will outlast it.

William James

1. Simplify: $\dfrac{-6+(-4)(3)}{(-2)(-3)(-3)}$.

2. Simplify: $\dfrac{18+(2)(-5)(-1)+1}{30+(-1)(29)}$.

3. Find the positive value of y if the point of coordinates (396, y) is on the graph of the parabola: $-x + y^2 = 4$.

4. Determine the y-intercept of the quadratic function: $f(x) = -2(x-3)^2 + 33$.

(1-29-20-15)

I prefer the errors of enthusiasm to the indifference of wisdom.

Anatole France

1. Simplify: $10 - 12 \div 4(-3) \div (3)(-3)$.

2. Simplify: $(-15)(-3) + 5(-3)$.

3. Determine the y-coordinate of the point of intersection of $f(x) = 3x - 1$ and $g(x) = x^2 - 29$ in the first quadrant.

4. Find half of the sum of the solutions of the quadratic equation: $2x(x - 30) = 7$.

(1-30-20-15)

Always bear in mind that your own resolution to success is more important than any other one thing.

Abraham Lincoln

1. Evaluate: $\dfrac{a^2 + 4b}{-3}$ for a = 3 and b = - 3.

2. Evaluate: $\dfrac{-x^2 - 3y}{-x+2y}$ for x = 5 and y = 2.

3. For what positive value of x is y = - 18, if y = - $|3 - 2x|$ + 19.

4. Find the smallest value of (x + y), if x − y = 3 and y = - $2\sqrt{x}$ + 12.

(1-31-20-15)

None are so old as those who have outlived enthusiasm.

Henry David Thoreau

1. Simplify: $(- 3)^0 + 2 - 1$.

2. Simplify: $- 5^2 + 4(7) - 2$.

3. Find x if g(x) = - 10 and g(x) = - $\sqrt[3]{x + 7}$ - 7.

4. Determine the largest value of x such that the point (x, -9) is on the graph of the function: H(x) = - $3|x - 8|$ + 12.

(2-1-20-15)

Success generally depends upon knowing how long it takes to succeed.

Charles de Montesquieu

1. Simplify: $23 + 4(-3) - 3^2$.

2. Evaluate: $\frac{-a+b}{15+2b}$ for a = - 12 and b = -6.

3. Let $f(x) = 2x - 3$ and $g(x) = x + 2$. Determine the value of x for which the composition of f and g is 41.

4. If $h(x) = x^2 - 10$ and $k(x) = 2x - 1$, evaluate h composed with k of -2.

(2-2-20-15)

If we were logical, the future would be bleak indeed. But we are more than logical. We are human beings, and we have faith, and we have hope, and we can work.

Jacques Cousteau

1. Evaluate: $(\frac{-b^2}{4a} + c)$ for a = 1, b = 4, and c = 6.

2. Simplify: $(-4)^2 + 20 \div 5(-4) + 3$.

3. If $f(x) = -3x + 2$ and $g(x) = -x^2 - 1$, evaluate $(f \cdot g)(2)$.

4. For the functions $f(x) = \sqrt{9 - x}$ and $g(x) = 2x^2 - 20$, evaluate $\frac{g}{f}(5)$.

(2-3-20-15)

Optimism is the faith that leads to achievement. Nothing can be done without hope or confidence.

Helen Keller

1. Solve the equation: $5(8 - 13) = x - 27$.

2. Solve the equation: $2x + 3(4 - 6) = 2$.

3. Let $f(x) = \sqrt{x - 4}$. Find $f^{-1}(4)$.

4. If $g^{-1}(x) = \dfrac{x^3}{27} - 30$, find the value of $g(x)$ when $x = 95$.

(2-4-20-15)

He who has a WHY to live for can bear almost any HOW.

Friedrich Nietzsche

1. Two angles are supplementary. One angle measures $(93 - x)$ and the other angle measures $(95 - 3x)$. Find the value of x.

2. One of two complementary angles is x and the other is………… $(15x + 10)$. Find the smaller angle.

3. One day Ana spent $5 more than she donated to charity. The difference between the square of her donation and five times the amount she spent is $125. How much did she spend?

4. In problem #3, find the amount Ana donated to charity.

(2-5-20-15)

The fishermen know that the sea is dangerous and the storm terrible, but they never found these dangers sufficient reason for remaining ashore.

Vincent van Gogh

1. Solve the equation: $6(5x) = \frac{75}{5}$. Then find 4 times the solution.

2. If you have 7 times as many oranges than pears and you have 42 oranges, how many pears do you have?

3. If $f^{-1}(x) = 3x^2 + 20$, find the left-side limit of the domain of $f(x)$.

4. If $k(x) = \sqrt[4]{60 - 4x}$, find the maximum value of $k^{-1}(x)$.

(2-6-20-15)

The secret of success in society is a certain heartiness and sympathy.

Ralph Waldo Emerson

1. If you have triple the number of dimes than quarters, and you have eight coins over all, how many quarters do you have?

2. Solve the equation: $8x + 5x = 91$.

3. Find the sum of the x- and y-coordinates of the vertex of the quadratic function: $f(x) = 3x^2 - 36x + 122$.

4. What is the y-intercept of the graph of the parabola:...............
   $G(x) = 2(x - 2)^2 + 7$?

(2-7-20-15)

Skill and confidence are an unconquered army.

George Herbert

1. If you subtract the double of a number from five times the number, the answer is 6. Find the number.

2. George has two times as many apples as pears. If he has 16 apples, how many pears does George have?

3. For the parabola $H(x) = -2x^2 + 5x + c$, find its y-intercept if the point $(5, -5)$ is on the graph of $H(x)$.

4. In the quadratic function $G(x) = a(x-3)^2 - 75$, find the value of a if its graph passes through the point $(-2, 300)$.

(2-8-20-15)

Is there anyone so wise as to learn by the experience of others?

Voltaire

1. A rectangle has an area of 24. If the length is 6, find the difference between the length and the width.

2. A square has the same area as a rectangle with length 27 and width 3. Find the side of the square.

3. A third degree polynomial function has the x-intercepts -2, 2, and 5. Find its y-intercept if the leading coefficient is 1.

4. Find the product of all the zeros of the function:
    $g(x) = x(x^2 + 23) - 3(3x^2 + 5)$.

(2-9-20-15)

What torments of grief you endured, from evils that never arrived.
Ralph Waldo Emerson

1. The formula $A = \frac{h}{2}(b + B)$ represents the area of a trapezoid. Find the height (h) if A = 20, b = 8, and B = 12.

2. Determine the difference between the length and the width of a rectangle if its area is 56 and the width is 4.

3. Find the largest value in the range of the polynomial function

   $F(x) = -3x^2 - 12x + 8.$

4. Determine the sum of all the zeros of the function
   $g(x) = x(x^2 + 68) - 3(5x^2 + 32).$

(2-10-20-15)

I am not afraid of tomorrow, for I have seen yesterday and I love today.
William Allen White

1. Simplify, and then evaluate the expression for
   x = 1: 5x(- 3x) + 17.

2. Simplify the expression $- 3x^2(x^2 + 2x) + 3x^2 - 1$, and then evaluate it for

   x = - 2.

3. Use synthetic division to find the remainder of
   $(x^2 + x + 14) / (x - 2).$

4. Use synthetic division to find g(- 1) if $g(x) = x^3 + 2x^2 + 3x + 17.$

(2-11-20-15)

A wise man will desire no more than what he may get justly, use soberly, distribute cheerfully, and leave contently.

Benjamin Franklin

1. The length and the width of a rectangle are respectively $(2x^2 + x)$ and $(4x)$. Find the difference between the length and the width if $x = 2$.

2. The volume of a rectangular solid is $V = lwh$. If $l = 4x + 2$, $w = x$, and $h = 2x$, find the volume when $x = 1$.

3. Use the long division of polynomials to find the remainder of the division: $(x^3 - 2x^2 - 4x + 28) / (x^2 - 4)$.

4. Use any method to determine the remainder of the division: $(x^3 + 2x^2 + x + 17) \div (x + 2)$.

(2-12-20-15)

Common sense is instinct. Enough of it is genius.

George Bernard Shaw

1. What is the smallest prime factor of 34?

2. Find the largest prime factor of 156.

3. Let $u = 2 - 5i$ and $v = 5 + 2i$. Write $u \cdot v$ in the standard complex form $(a + bi)$. What is the value of a?

4. With u and v from problem #3, simplify $v(-u) - 6i$. Write the answer in $a + bi$ form and determine the value of $(b + 2)$.

(2-13-20-17)

There is a place deep within us that wants to know that our life has made a difference.

Wayne Dyer

1. Write 150 in prime factorization using exponents. What is the exponent of the largest prime factor?

2. Write 350 in prime factorization and find the product of the smallest and the largest prime factors other than 1.

3. Let $z = x + yi$ be a complex number and v its complex conjugate. Find $|x \cdot y|$ if z·v = 41 and x = 4.

4. Rationalize the denominator of the fraction: $\frac{1}{3i + \sqrt{8}}$. What is the new denominator in positive form?

(2-14-20-17)

I long to accomplish a great and noble task, but it is my chief duty to accomplish small tasks as if they were great and noble.

Helen Keller

1. In a group of 39 people $\frac{4}{6}$ are women. How many times is the number of women larger than the number of men?

2. If there are four times as many men as women in a group of 75 people, how many women are there?

3. If x = 2i is a complex zero of $f(x) = x^4 - 9x^3 + 24x^2 - 36x + 80$, find the product of the real zeros of f(x).

4. Let $g(x) = x^3 - 23x^2 + 142x - 330$. Find the real zero of g(x), if....... $(4 + i\sqrt{6})$ is a complex solution of the equation g(x) = 0.

(2-15-20-15)

True success is overcoming the fear of being unsuccessful.
Paul Sweeney

1. Find the numerator in the mixed number form of the fraction $\frac{67}{5}$.

2. What is the numerator of the fraction in simplest form equivalent to $3\frac{1}{5}$.

3. Let $f(x) = 2x^2 + ax + 5$. Find the real value of 'a' if $x = -5 + \frac{3}{2}\sqrt{10}$ is a zero of the function $f(x)$.

4. If $g(t) = -t^3 + at^2 + bt - 14$, and $t = 3 - \sqrt{23}$ is a t-intercept of $g(t)$, find the value of $(a + b + 2)$.

(2-16-20-17)

To be on the alert is to live; to be lulled into security is to die.
Oscar Wilde

1. Find the numerator of the fraction $\frac{74}{111}$ written in simplest form.

2. Find the denominator of the fraction $\frac{30a^3b}{34ba^2}$ written in simplest form.

3. For the rational function $g(x) = \frac{3x^2 - 4x + b}{x - 2}$,

   find the value of b such that $g(-2) = -10$.

4. Find the positive t-intercept of the rational function $f(t) = \frac{t^2 - 12t - 45}{t + 15}$.

(2-17-20-15)

The value of life lies not in the length of days, but in the use we make of them; a man may live long yet live very little.

Michel de Montaigne

1. Simplify the fraction $\dfrac{-36x^2yz^4}{18xyz^2}$, then indicate the exponent of z.

2. Simplify the fraction $\dfrac{-27x^3}{3x^2}$ and evaluate your answer for x = - 2.

3. The rational function f(x) = $\dfrac{x-3}{x^2-x-c}$ has two vertical asymptotes whose x-intercepts have a difference of 9. Find the value of c, then graph the function.

4. For the rational function g(x) = $\dfrac{x^2+4x+3}{x-11}$ find the y-intercept of the slant (oblique) asymptote, then graph the function.

(2-18-20-15)

Hope is one of the principal springs that keep mankind in motion.

Thomas Fuller

1. Find the unit rate in dollars per pound if you buy 24lb of rice for $48.

2. What is the unit rate in miles per hour if you cover 57 miles in 3 hours?

3. The area of a rectangular garden is 360ft². If one side is 2ft shorter than the other, find the longer side.

4. The perimeter of a rectangular flower garden is 60ft. Out of all possible such gardens, find the longer side of the garden with the greatest area.

(2-19-20-15)

Man is not born to solve the problems of the universe, but to find out what he has to do ... within the limits of his comprehension.

Johann von Goethe

1. Find the value of x in the proportion: $\dfrac{x}{62} = \dfrac{3}{93}$.

2. Solve the proportion for t: $\dfrac{35}{t} = \dfrac{7}{4}$.

3. Find b if the point of coordinates (4, b) is on the graph of the exponential function $f(x) = 2^x + 4$.

4. Determine the value of c such that the point (0, 16) is on the graph of $y = 3^{-x} + c$.

(2-20-20-15)

We challenge one another to be funnier and smarter ... It's the way friends make love to one another.

Annie Gottlieb

1. Find the value of w in the proportion: $\dfrac{17}{51} = \dfrac{w}{6}$.

2. Solve the proportion for z: $\dfrac{6}{7} = \dfrac{18}{z}$.

3. If 'a' is the largest solution of $\dfrac{1}{x-1} + \dfrac{1}{x+1} \geq \dfrac{3}{4}$, evaluate 6a + 2.

4. Find the maximum of the function $f(x) = -\sqrt{x-2} + 17$, on the interval $[6, +\infty)$.

(2-21-20-15)

To be what we are, and to become what we are capable of becoming,
is the only end of life.

Robert Louis Stevenson

1. Perform the operation and simplify. What is the value of the numerator?

$$\frac{5}{6} - \frac{1}{4} + \frac{1}{12}.$$

2. Perform the operation and simplify. Find the value of the numerator.

$$\frac{1}{7} + \left(\frac{-1}{29}\right).$$

3. Find the y-intercept of the exponential function:
    $g(x) = 3^{-2x} + 17.$

4. The point $(b, \frac{1}{2})$ is on the graph of $f(x) = 2^{-x+13} + \frac{1}{4}$. Find $(b + 2)$.

(2-22-20-17)

Education should be the process of helping everyone to discover his
uniqueness.

Leo Buscaglia

1. Simplify and identify the first digit of the numerator:

$$\frac{-3}{16} + \frac{9}{20}$$

2. Add, simplify, and indicate the numerator of your answer:

$$\frac{4}{15} + \frac{3}{6}$$

3. Find the x-intercept of the function: $f(x) = \log_2(x - 4) - 4.$

4. If the point $(100, b)$ is on the graph of $h(x) = -\log x + 17$, find
    ...$(b + 2)$

(2-23-20-17)

All happiness depends on courage and work.

Honore de Balzac

1. Perform the operation and simplify: $2\frac{5}{7} - \frac{1}{7} - \frac{8}{14}$.

2. Simplify in improper fraction form and identify the numerator: $2\frac{3}{5} + 2\frac{2}{10}$.

3. Evaluate the function for x = 30: $k(x) = \log_5(x - 5) + 18$.

4. If $f(x) = \ln(x + e^5)$, simplify f(x) for $x = e^3(1 - e^2)$, and find the product of your answer and the y-intercept of f(x).

(2-24-20-15)

Our remedies oft in ourselves do lie.

William Shakespeare

1. Subtract, write your final answer in mixed-number form, and state the integer part of your answer: $4\frac{1}{4} - 1\frac{1}{2}$.

2. Identify the denominator of your answer if you divide $5\frac{3}{5}$ by $1\frac{2}{3}$.

3. Evaluate the expression for x = 27:

$\ln e^{13} + \log_3 x + \log_3(x + 54)$.

4. Find the product of x and y if: $\log(xy\sqrt[3]{x - 2}) = \frac{\ln 15}{\ln 10}$ and x = 3.

(2-25-20-15)

Happy the man who has broken the chains which hurt the mind, and has given up worrying, once and for all.

Ovid

1. Simplify: $-\left(\frac{4}{3}\right)^2\left(-\frac{3^2}{2^3}\right)$.

2. Simplify and indicate the numerator of your answer:

   $\left(-\frac{13}{8}\right)\left(-\frac{2^4}{3^2}\right)$.

3. Condense the expression and evaluate it for $x = 2^8$:
   $\log_2 x + 2\log_2(4x) - \log_2(2x) + \log_x x$.

4. Expand the expression $\log(x\sqrt[3]{x^2})$ and evaluate your answer for $x = 10^9$.

(2-26-20-15)

If you keep saying things are going to be bad, you have a good chance of being a prophet.

Isaac Bashevis Singer

1. Multiply and simplify. Determine the numerator of your answer:
   $\left(-\frac{1}{5}\right)^3\left(-\frac{2}{3}\right)^2\left(-\frac{5^2}{2}\right)$.

2. Simplify and find the denominator of your answer:
   $\left[\frac{1}{9}+\frac{2}{3}\right] \div \left[3\frac{1}{3} - \left(\frac{5}{6} - \frac{1}{2}\right)\right]$.

3. Condense and simplify. Then, evaluate the expression for $x = e$:
   $\ln x^5 + 6\cdot \ln x^2 + \ln x^3$.

4. Find the value of $(e^x + 2)$ if: $e^{2x} = 12e^x + 45$.

(2-27-20-17)

Faced with crisis, the man of character falls back on himself.

Charles de Gaulle

1. Divide and simplify: $(\frac{5}{6})^2 \div [\frac{1}{8}(\frac{5}{3})^2]$.

2. Simplify and find the numerator of your final answer:
   $\frac{7}{5} \div [\frac{1}{2} - (\frac{1}{2})^2]$.

3. Simplify the expression and evaluate the answer for $x = 10$:

$$\frac{\log x^{\ln e^4}}{\log_{x^5} 10}.$$

4. Find the value of $\log_3 x$ if: $\log_3 x^2 = \log_3 x^4 - 30$.

(2-28-20-15)

The people who get on in this world are the people who get up and look for the circumstances they want, and, if they can't find them, make them.

George Bernard Shaw

1. Simplify: $\dfrac{2(\frac{-1}{3})^2}{\frac{1}{9} - (\frac{1}{3})^3}$.

2. Simplify: $\dfrac{-16+(-3)(-5)}{-(-5)^2 - 3(-2)^3}$.

3. Find the value of e^x if: $38e^x + 40 = 2e^{2x}$.

4. Evaluate $(5|x| + 2)$, if x is a solution of the equation:
   $\log_3(3x^2) - 3 = 0$.

(3-1-20-17)

Even if it's a little thing, do something for those who have need of help, something for which you get no pay but the privilege of doing it.

Albert Schweitzer

1. You want to build two book-cases, each with three shelves, each shelf being 3 feet long. The wood is sold in 6-foot long boards. How many boards do you need?

2. If the boards in problem #1 are 10-foot long, how many feet of wood do you have left if you buy most economically?

3. Evaluate (2x) if x is the solution of the equation: $\log x^2 + 1 = 3\log x$.

4. After how many months (rounded to the nearest whole number of months) will your simple interest be $31.56 (rounded to the nearest cent), if you invest $500 at 5% annual interest compounded annually?

(3-2-20-15)

No man can produce great things, who is not thoroughly sincere in dealing with himself.

James Russell Lowell

1. Solve the equation: $\dfrac{x}{-5} = -\dfrac{9}{15}$.

2. Find the solution of the equation: $\dfrac{-5}{y} = \dfrac{15}{-9}$.

3. Find the largest value of y that satisfies the system of equations:
$$-2x + y = 10 \qquad\qquad x^2 - y = 5$$

4. Find the value of x that satisfies the system of equations when y is negative:
$$x - y^2 = -1$$
$$x + y = 11$$

(3-3-20-15)

Ask yourself the secret of your success. Listen to your answer, and practice it.

Richard Bach

1. Solve the equation: $-\frac{3}{4}t = -\frac{18}{8}$.

2. Solve: $\frac{z}{3^3} = \frac{13}{27} - \frac{1}{3}$.

3. Solve the system of equations and determine the value of x:

   $x + y - z = 6$
   $2x + y - 2z = 11$
   $-x + 5y + 2z = 15$

4. In problem #3, determine the value of $(z + 2)$.

(3-4-20-17)

How poor are they that have not patience? What wound did ever heal but by degrees?

William Shakespeare

1. Solve for x : $\frac{x}{2^4} = \frac{1}{8} + \frac{1}{16}$.

2. Perform the operations and simplify:
   $7x - (3x^2 - 5) + 3(x^2 + x) - 10x$.

3. Solve the system of linear equations and indicate the value of b:

   $5a + 3c = b + 15$
   $2b = a + 2c + 12$
   $a + b = c + 3$

4. In problem #3 find the value of $(c + 2)$.

(3-5-20-17)

Be wary of the man who urges an action in which he himself incurs no risk.

Joaquin Setanti

1. Perform all operations and simplify: $-2(x + 4) - 6(2x - 1) + 5 + 7(2x)$.

2. Subtract $(3x^2 - 4x - 2)$ from $(5x^2 + x + 6)$, then evaluate your answer for $x = -2$.

3. Evaluate the 3×3 determinant:
$$\begin{vmatrix} 2 & 1 & 0 \\ 3 & 2 & 1 \\ 17 & 0 & 3 \end{vmatrix}$$

4. Find x such that the value of the following determinant is 42.
$$\begin{vmatrix} 3 & -2 \\ x & 4 \end{vmatrix}$$

(3-6-20-15)

In times like these, it helps to recall that there have <u>always</u> been times like these.

Paul Harvey

1. Subtract $(-3x^2 + 4)$ from $(x^2 + 3x - 15)$, then evaluate your answer for $x = 2$.

2. Multiply and simplify, then evaluate your answer for $x = -2$: $(2x - 3)(2x + 3)$.

3. Find the length of the interval on the real x-axis that defines the domain of the function $F(x) = x^3 - \sqrt{100 - x^2}$.

4. Find the positive x-coordinate of the point of intersection between the line passing through $(4, -3)$ and $(-4, 5)$, and the parabola $y = x^2 - 239$.

(3-7-20-15)

If your daily life seems poor, do not blame it; blame yourself, tell yourself
that you are not poet enough to call forth its riches.

Rainer Maria Rilke

1. Multiply and simplify your answer,
   then evaluate it for x = 2: $(x - 1)(7 - x^2)$.

2. The first of three books cost $2 more than the second. The third
   cost $3 less than the second. How much does the second book
   cost if the first and the third together cost $15?

3. Find the product of the x- and the y-coordinates of the center of
   the circle of equation: $x^2 + y^2 = 8x + 10y + 50$.

4. Write a linear function that expresses the sale price of an item
   that is discounted by 10% from the list price. Find the list price if
   the sale price of the item is $13.50.

(3-8-20-15)

Inspirations never go in for long engagements; they demand immediate
marriage to action.

Brendan Francis

1. A flower garden is twice as long as it is wide. The sum of the
   length and the width is 9. Find the width of the garden.

2. Solve the equation: $\dfrac{-72}{4} = -8x - (-6x)$.

3. The piece function g(x) is defined as $g(x) = 3x - 5$ if $x \le 6$, and
   $g(x) = -x^2 + 84$ if $x > 6$. Find g(8).

4. If $f(x) = -2x + 23$ and $g(x) = x^2$, evaluate $[(f \circ g)(-2) + 2]$.

(3-9-20-17)

Sadness flies on the wings of the morning, and out of the heart of darkness comes the light.

Jean Giraudoux

1. Solve: $\frac{4}{3}y = 4^2 - 12$.

2. Solve: $-5 + 2x + 3x = -4 + 6x - 11$.

3. Find the product of the upper and the lower limits of the domain of the real function: $f(x) = \sqrt[3]{x} - \sqrt[4]{9x - 20 - x^2}$.

4. Find the maximum of: $g(x) = -2x^2 - 12x - 3$.

(3-10-20-15)

There is nothing so wretched or foolish as to anticipate misfortunes. What madness is it in expecting evil before it arrives?

Marcus Annaeus Seneca

1. Solve: $2(x - 3) - 3(x + 1) = -4x$.

2. Solve: $(5x^2 - 2x - 3) - (x - 11) = 5(x^2 - 5)$.

3. Find the product of the x- and the y-coordinates of the point of symmetry for the graph of: $x^2 + y^2 - 8x - 10y + 32 = 0$.

4. Determine the x-coordinate of the highest point on the graph of:
$$h(x) = -\frac{1}{3}x^2 + 10x - 73.$$

(3-11-20-15)

*Friendship with oneself is all-important, because without it one cannot
be friends with anyone else.*

Eleanor Roosevelt

1. Solve: $4x^2(3x^3 - 2) - 3x = 2x(6x^4 - 4x) - 9$.

2. Solve: $\frac{1}{2}x + \frac{1}{4} = \frac{25}{4}$.

3. For $x = 7$, find the y-coordinate of the point on the oblique
    (slant) asymptote of the function $g(x) = \frac{3x^2 - 4x + 5}{x - 1}$, then sketch
    the graph of $g(x)$.

4. Let P be a point on the graph of $f(x) = x^2 - 9$. Find the distance
    from point P to the x-axis when $x = 2\sqrt{6}$.

(3-12-20-15)

*Conditions are never just right. People who delay action until all factors
are favorable do nothing.*

William Feather

1. Solve: $2x + 1\frac{1}{3} = \frac{22}{3}$.

2. The length of a rectangle is $(x - 6)$ and the width is 5.
    Find x if the perimeter is 24.

3. If you invest an amount P at 5% interest compounded semian-
    nually, at the end of the year your account balance is $22.05.
    Find P.

4. Find the real zero of the function with real coefficients
    $f(x) = x^3 - 17x^2 + 33x - 45$, given that $x = 1 - i\sqrt{2}$ is a zero of $f(x)$.

(3-13-20-15)

I am in the present. I cannot know what tomorrow will bring forth. I can know only what the truth is for me today. That is what I am called upon to serve.

Igor Stravinsky

1. Jon has 2 more apples than Mary, and Tom has one less apple than Mary. All three together have 10 apples. How many apples does Mary have?

2. The first side of a triangular flower garden is x. The second is 4ft longer than the first, and the third is double the first. Find the length of the first side if the perimeter of the garden is 60ft.

3. Use Cramer's rule (with determinants) to find y from the system:

$$2x - y = 10$$
$$x + 2y = 55.$$

4. Use Gauss-Jordan elimination (with matrices) to solve the system from problem #3 for x.

(3-14-20-15)

To accept whatever comes, regardless of the consequences, is to be unafraid.

John Cage

1. Write the number 7.28 in fraction form, simplify the fraction, then round the denominator to the tens position. What is the tens digit?

2. Write the decimal number 0.6 in fraction form, simplify the fraction, then determine the product of the numerator and the denominator.

3. The base of a pyramid is a square with the diagonal $6\sqrt{2}$. Find the height of the pyramid if its volume is 240.

4. If the volume of a sphere is 4500π, find its radius.

(3-15-20-15)

When I first open my eyes upon the morning meadows and look out upon the beautiful world, I thank God I am alive.
Ralph Waldo Emerson

1. Write $\frac{349}{100}$ in decimal form and round the answer to a whole number.

2. Write the fraction $\frac{135}{8}$ in decimal form and round it to tenths. What is the integer part of your answer?

3. A sphere is inscribed in a cube. If the volume of the sphere is $\frac{4000}{3}\pi$, find the side of the cube.

4. Triangle ABC is a 30-60-90 right triangle with <B = 90°. If the area of the triangle is $\frac{225\sqrt{3}}{2}$, find the length of the median BD.

(3-16-20-13)

Success based on anything but internal fulfillment is bound to be empty.
Dr. Martha Friedman

1. Simplify by combining like terms:
   7.4x − 3.2y − 2(3.7x − 1.6y − 1.5).

2. Combine like terms and find the coefficient of t:
   23.5t − 4.2z − (6.5t − 3.2z).

3. In triangle ABC, points D and E are on AB and AC respectively, such that DE is parallel to BC. If DE = 5, BC = 25, and BD = 16, find AB.

4. The perimeter of a right triangle is $10(2 + \sqrt{2})$. Find the hypotenuse if one leg is 5 and the area of the triangle is $25\sqrt{2}$.

(3-17-20-15)

If you've had a good time playing the game, you're a winner even if you lose.

Malcolm Forbes

1. Multiply (1.76)(1.7) by hand and round your answer to a whole number.

2. Divide (92.82) by (5.1) by hand and determine the integer part of your answer.

3. The volume of a cube is $\frac{64000\sqrt{3}}{9}$. Find the radius of the sphere circumscribed to the cube.

4. In the equilateral triangle ABC the altitudes from A and B intersect at D. Find the height of the triangle if BD is 10.

(3-18-20-15)

Go ahead with your life, your plans ... Don't waste time by stopping before the interruptions have started.

Richard L. Evans

1. A round-trip bus fare to work is $1.85. A bus pass for a 20-day work month is $34. How much cheaper is to buy the pass?

2. Right now Johnny is 13.5 years old. When he will double his age his sister will be 46 years old. How much older than him will she be then?

3. Triangles ABC and MNP are similar in this order of the vertices. If their perimeters are 34 and 68 respectively, and AB = 10, find MN.

4. The area of a trapezoid is 375. The lengths of the parallel sides are 20 and 30. Find the height of the trapezoid.

(3-19-20-15)

One should ... be able to see things as hopeless and yet be determined to make them otherwise.

F. Scott Fitzgerald

1. What percent does the fraction $\frac{12}{400}$ represent?

2. 3.7 is what percent of 18.5?

3. Find the area of a triangle if the coordinates of its vertices are (2, 2), (- 4, -2), and (6, -2).

4. A triangle has the vertices (1, 7), (9, 1), and (10, 4). Find its area.

(3-20-20-15)

Trigonometry (3, 4)

Your real security is yourself. You know you can do it, and they can't ever take that away from you.

Mae West

1. Estimate 20% of 15.5 to a whole number.

2. Evaluate 40% of 52.5.

3. Find the smallest positive angle co-terminal with 380°.

4. Find the positive angle in the first quadrant co-terminal with (- 705°).

(3-21-20-15)

Everything in life that we really accept undergoes a change. So suffering must become love. That is the mystery.
Katherine Mansfield

1. By how much more of a percent is 26 out of 200, compared to 5.6 out of 56?

2. What is the original price of an item discounted by 15% to a sale price of $18.7?

3. A pulley does 240 rotations per minute. Through how many degrees does a point on the edge of the pulley move in $\frac{1}{72}$ seconds?

4. Convert the decimal degree angle 8.26° to degrees, minutes, and seconds. What is the number of minutes?

(3-22-20-15)

Shallow men believe in luck, wise and strong men in cause and effect.
Ralph Waldo Emerson

1. What percent of 800 is 24?

2. Find the mean of the set of data: {9, 21, 14, 32, 25, 37}.

3. Two vertical angles are given by (3x − 10) and (5x − 50). Find the value of x.

4. The three angles of a triangle are 3x, 4x, and 5x. Find the value of (x + 2).

(3-23-20-17)

We create our fate every day…most of the ills we suffer from are directly traceable to our own behavior.

Henry Miller

1. If 9.6 is 320% of x, find the value of x.

2. Find the median of the set of data: {3, 30, 40, 15, 49, 18}.

3. Triangles ABC and MNP are similar in this order of the vertices. If PN = 12 and the perimeter of the triangles are 45 and 27 respectively, find BC.

4. If a 30ft tall tree casts a shadow of 50ft, how long would the shadow of a 9ft tall tree be at the same time?

(3-24-20-15)

We should every night call ourselves to an account: What infirmity have I mastered today? What passions opposed? What temptation resisted? What virtue acquired?

Marcus Annaeus Seneca

1. How many feet are there in 36 inches?

2. How many yards are there in 75 feet?

3. Simplify the expression: 40 sin 390°.

4. Simplify the expression: 30 cos 420° + 2.

(3-25-20-17)

There is change in all things. You yourself are subject to continual change and some decay, and this is common to the entire universe.

Marcus Aurelius

1. How many meters are there in 300cm?

2. Find the area (in square hectometers) of a rectangular ranch if its dimensions are 200m by 1300m.

3. The point (- 2, - 2) is on the terminal side of angle θ . Evaluate the expression: $- 10\sqrt{2} \sin \theta - 10\sqrt{2} \cos \theta$.

4. The point (0, - 5) is on the terminal side of angle β. Evaluate: $13 \cos \beta - 15 \sin \beta$.

(3-26-20-15)

Every man has the right to risk his own life in order to preserve it. Has it ever been said that a man who throws himself out the window to escape a fire is guilty of suicide?

Jean Jacques Rousseau

1. Angles (x + 37) and (3x + 41) are complementary. Find x.

2. The first of two supplementary angles is (2x + 80) and the second is (x + 19). Find x.

3. The terminal side of angle ∂ in standard position is on the line of equation: x + y = 0. Evaluate the expression: $- 13\sqrt{2} \cos \partial + 7\sqrt{2} \sin \partial$.

4. If the terminal side of angle α in standard position is on the line of equation: x − 2y = 0, evaluate the expression: (5sin α) (7.5cos α) + 2.

(3-27-20-17)

I steer my bark with hope in my heart, leaving fear astern.
Thomas Jefferson

1. Two parallel lines are cut by a transversal. If one of the obtuse angles created is equal to 120°, find $\frac{1}{20}$ of the acute angle.

2. Two parallel lines are cut by a transversal such that any two adjacent angles are (2x + 4) and (2x + 64). Find x.

3. Evaluate: $12\sqrt{3}$ tan 60° + 16 cot 135°.

4. Evaluate the expression: 2 + 10 sec 300° + 5 cot 315°.

(3-28-20-17)

Elementary Algebra (1, 2)

It is only possible to live happily-ever-after on a day-to-day basis.
Margaret Bonnano

1. Evaluate the expression: $\frac{2m-3n}{6}$ for m = 6 and n = - 2.

2. Evaluate $\frac{-3x+4y}{2}$ when x = - 10 and y = 7.

3. One of the acute angles of a right triangle is 30°. Use trigonometric functions to find the hypotenuse if the opposite side of the 30° angle is 10.

4. Use trigonometric functions to find the leg of an isosceles right triangle if the hypotenuse is $15\sqrt{2}$.

(3-29-20-15)

*The man is happiest who lives from day to day and asks no more, gar-
nering the simple goodness of life.*

Euripides

1. Evaluate: $\frac{-5x-y}{4}$ for x = -3 and y = 3.

2. Use the distributive property to factor: 24x + 18y + 6. Then evaluate the expression in parentheses for x = 5 and y = 3.

3. Find the smallest positive angle x such that: $\sin \frac{9}{2}x - 1 = 0$.

4. Find the smallest positive angle x such that: cos 4x - .5 = 0.

(3-30-20-15)

Life shrinks or expands in proportion to one's courage.

Anais Nin

1. Factor the expression using the distributive property: $12xy - 8x^2$
   + 4x. How many terms are there left in the parentheses?

2. Perform the operations and simplify: $\frac{3}{7} + \frac{5}{21} + \frac{1}{14}$. What is the numerator of your answer?

3. Find the acute angle x larger than 10 and less than 30 such that:
   $\sin \frac{45}{4}x = \cos \frac{45}{4}x$.

4. Find x in the first quadrant such that: tan 2x = cot 4x.

(3-31-20-15)

Great men are they who see that the spiritual is stronger than any material force, that thoughts rule the world.

Ralph Waldo Emerson

1. Simplify: $\dfrac{9}{4} \div \dfrac{36}{64}$.

2. Simplify: $\dfrac{\frac{3}{5}}{\frac{12}{20}}$.

3. Find the reference angle for: (- 200°).

4. Find the reference angle for : (345°).

(4-1-20-15)

If I can stop one heart from breaking, I shall not live in vain.

Emily Dickinson

1. How many rational numbers are there in the set:
   $\{- 3, \frac{4}{5}, \pi, \sqrt{3}, 0.2, \sqrt[3]{9}, \sqrt{25}\}$.

2. How many irrational numbers are there in this set:
   $\{-\frac{1}{5}, 0.3, \sqrt[5]{5}, - 1, - \pi\}$.

3. Evaluate: - 40 cos (- 240°).

4. Evaluate: $2 + 5\sqrt{3} \cot 210°$.

(4-2-20-17)

Strong hope is a much greater stimulant of life than any single realized joy could be.

Friedrich Nietzsche

1. Combine like terms, then evaluate your answer for
   x = - 3 and y = 4:
   -5x + 4y − 2x − 8y − 1.

2. Simplify by combining like terms, then evaluate the answer for
   a = -1, b = -5, and c = 4:
   3a − 7 + 5b − 7a − 3b + 4c.

3. Evaluate the expression and give the exact value:
   $8 \cos^2 60° - 18\sqrt{2} \sin(-45°)$.

4. Find the exact value of the expression:
   $6 \sec^2 30° + 2 + 7\sqrt{3} \cot 60°$.

(4-3-20-17)

Some people are making such thorough preparation for rainy days that they aren't enjoying today's sunshine.

William Feather

1. Perform the operations and simplify: (- 3)(- 4)÷(-3)(- 1).

2. Simplify: $\dfrac{6(-13)(-4)}{(-3)(-26)}$.

3. Find the exact value of the expression:
   $8\cos^2 135° + 9\csc^2 225° - 2\cot^2 315°$.

4. Find the exact value in simplest form for the expression:
   $7(\sin^2 30° + \cos^2 390°) + 4(\tan^2 45° + \cot^2 135°) + 2$.

(4-4-20-17)

There is no such thing as a long piece of work, except one that you dare not start.

Charles Baudelaire

1. Simplify the expression: $\dfrac{[(-5)(-5)-2(5)]2^2}{(-5)(-3)(-2)^0}$.

2. Simplify: $\dfrac{[-4(4)+4(2^3)](8-3)}{|-2^4|}$.

3. Find the exact value of 'a' such that:
   $8\cos^2 30° + a{\cdot}\cos^2 60° + \cot^2 45° = 12$.

4. Find the precise value of 'b' that satisfies the equation:
   $3\sec^2 60° + b{\cdot}\tan^2 135° = 27$.

(4-5-20-15)

We secure our friends not by accepting favors but by doing them.

Thucydides

1. Simplify and evaluate the answer for
   $x = 2$: $3(2x-5) - [2(1-2x)-1]$.

2. Simplify the expression, then evaluate your answer for $x = -4$:
   $-2(3-x) + 2[3-(2x+1)]$.

3. Find the smallest positive value of x if: $\cot^2(x+40°) - \dfrac{1}{3} = 0$.

4. Find the value of angle t in the first quadrant such that:
   $\sin^2(60° - t) = \dfrac{1}{2}$.

(4-6-20-15)

If you want to be respected by others, the great thing is to respect your-self.

Fyodor Dostoyevsky

1. Solve the equation: $\frac{1}{5} + x = 1 + \frac{16}{5}$.

2. Solve the equation: $-\frac{1}{6} = \frac{x}{-42}$.

3. The adjacent angle to the leg b of a right triangle is 30°. Use a trigonometric function to find the length of the hypotenuse if $b = 10\sqrt{3}$.

4. The hypotenuse of an isosceles right triangle is $5\sqrt{6}$. Find the exact value of $\frac{2}{5}$ of the area of the triangle.

(4-7-20-15)

I get up and I bless the light thin clouds and the first twittering of birds, and the breathing air and smiling face of the hills.

Giacomo Leopardi

1. Solve the equation: $-6 = -\frac{3}{2}y$.

2. If the length of a rectangle is $\frac{4}{3}$ of the width and the perimeter of the rectangle is 28, find the length.

3. The height of an equilateral triangle is $10\sqrt{3}$. Use trigonometric functions to find the side of the triangle.

4. Evaluate the expression precisely:
$2 + 20\sin^2 30° + 12\sin^2 60° + \sin^2 90°$.

(4-8-20-17)

*You can't measure time in days the way you can money in dollars,
because each day is different.*

Phillip Hewett

1. Solve the equation: $\frac{2}{3}(6-t)+\frac{1}{6}=\frac{3}{2}$.

2. Solve the equation $F=\frac{9}{5}C+32$ for C and write your solution in one fraction form. What is the denominator of your answer?

3. Angles a, b, and c are the smallest consecutive positive multiples of 30°. Evaluate: $30\sin a + 10\cos b + \frac{25}{2}\cot c$.

4. Angle x is an acute angle in an isosceles right triangle. Evaluate the expression: $7\tan x + 8\cot(90° - x) + \cos(180° - 2x) + 2$.

(4-9-20-17)

I was seldom able to see an opportunity until it had ceased to be one.

Mark Twain

1. The lengths of the two longer sides of a triangle are consecutive multiples of the shorter side. The perimeter of the triangle is 24. Find the shorter side.

2. In a trapezoid, the height is 6 and one of the parallel sides is 8. Find the other parallel side if the area of the trapezoid is 54.

3. How many degrees are there in an angle of $\frac{\pi}{9}$ radians?

4. Convert $\frac{\pi}{12}$ radians into degrees.

(4-10-20-15)

It is a common experience that a problem difficult at night is resolved in the morning after the committee of sleep has worked on it.

John Steinbeck

1. Find n if 175% of n is equal to 7.

2. What percent of 45 is equal to $4\frac{19}{20}$?

3. How many radians are there in $\left(\frac{3600}{\pi}\right)$ degrees?

4. Convert $\left(\frac{2700}{\pi}\right)$ degrees into radians.

(4-11-20-15)

A wise man fights to win, but he is twice a fool who has no plan for possible defeat.

Louis L'Amour

1. Including the tip, you paid for a meal $24. Considering you left a 20% tip, find the amount of the tip.

2. If 35% of a number is 4.2, find the number.

3. Evaluate the expression (angles are expressed in radians):
   $- 15 \cos \frac{\pi}{2} + 40 \sin \frac{\pi}{6}$.

4. Evaluate the expression (angles are expressed in radians):
   $3\sqrt{3} \csc \frac{\pi}{3} - 9 \cot \frac{7\pi}{4}$.

(4-12-20-15)

He who reigns within himself and rules his passions, desires, and fears is more than a king.

John Milton

1. The fence all around a rectangular vegetable garden is 22ft long. If the length of the garden is 3ft longer than the width, find the width.

2. If the perimeter of a rectangle is 48, find its longer side knowing that the two sides are consecutive odd integers.

3. Find the radius of a circle if the length of an arc of the circle intercepted by a central angle of 60° is $6\frac{2}{3}\pi$.

4. Find the radius of a circle if the area of a sector of central angle of 45° is $28\frac{1}{8}\pi$.

(4-13-20-13)

A cheerful frame of mind, reinforced by relaxation, which in itself banishes fatigue, is the medicine that puts all ghosts of fear on the run.

George Matthew Adams

1. Find the lower limit of the solution interval for the inequality: $3(x-2) \geq 10 - x$.

2. Determine the upper limit of the solution interval for the inequality:
$4(2x-5) - 42 \leq 5(x-4)$.

3. Evaluate the expression using the exact circular function value:
$-\frac{20\sqrt{3}}{3} \tan\left(-\frac{7\pi}{3}\right)$.

4. Evaluate the expression using the exact circular function value:
$-5 \tan\left(\frac{-5\pi}{4}\right) - 5 \sec\frac{8\pi}{3}$.

(4-14-20-15)

What a man thinks of himself, that is what determines, or rather indicate his fate.

Henry David Thoreau

1. Find the value of b if the points (-1, 2) and (b, 2) are 5 units apart.

2. Plot the points (2, - 8) and (2, 7) on a system of axes and find the distance between them.

3. Find the precise value of x in the first quadrant such that: $\sin 3x = \frac{\sqrt{3}}{2}$.

4. If $\cos 3t = \frac{\sqrt{2}}{2}$, what is the precise value of t in the first quadrant?

(4-15-20-15)

It's not our disadvantages or shortcomings that are ridiculous, but rather the studious way we try to hide them, and our desire to act as if they did not exist.

Giacomo Leopardi

1. Graph the linear equation and find the y-intercept:
   $3x -2y + 8 = 0$.

2. Graph the linear equation and find the x-intercept:
   $- 2x + 5y + 32 = 0$.

3. Find x° in the first quadrant such that: $\csc \frac{3x}{2} = 2$.

4. If $\cot 6x = 0$, what is the precise value of x° in the first quadrant?

(4-16-20-15)

*The young do not know enough to be prudent, and therefore they at-
tempt the impossible – and achieve it, generation after generation.*
Pearl S. Buck

1. If a and b are respectively the x- and the y-intercepts of the line:
   $3y = -7x + 21$, find the value of $(-a + b)$.

2. In the linear equation $ax + by = 34$, find the y-intercept when b
   is 2.

3. In the angular velocity equation (radians per second), find the
   time if: $\omega = \frac{5\pi}{12}$ and $\theta = \frac{25\pi}{3}$.

4. In the linear velocity equation (radians per second), find the
   time if $v = \frac{\pi}{9}$, r = 5, and $\theta = \frac{\pi}{3}$.

(4-17-20-15)

*Have no fear of change, as such and, on the other hand, no liking for it
merely for its own sake.*
Robert Moses

1. Graph the line and find its x-intercept: $\frac{3}{2}x - 2y = 6$.

2. Find the y-intercept of the line, then graph the line: $2y - 32 = 4$.

3. Find the amplitude of the trigonometric function:
   $y = 5(1 - 4\sin 3x)$.

4. Determine the period of the trigonometric function:
   $y = 2 - 3\cos\frac{2\pi}{15}(x - \frac{\pi}{6})$.

(4-18-29-15)

Every stage of life has its troubles, and no man is content with his own age.

Ausonius

1. Graph the linear equation and find the x-intercept: $6 - 3x = -6$.

2. Find the value of b such that the slope of the line passing through the points $(7, b)$ and $(-1, -5)$ is 3.

3. Find the phase shift in degrees of the function: $y = 2 + \cos(3x - \frac{\pi}{3})$, then graph the function.

4. Graph the trigonometric function over a one-period interval and determine the y-intercept in degrees: $y = -3(\tan x - 5)$.

(4-19-20-15)

All things are possible until they are proved impossible – and even the impossible may only be so as of now.

Pearl S. Buck

1. If the slope of a line is -4, find t such that the points $(-7, 12)$ and $(-5, t)$ are on the graph of the line. Then graph the line by plotting the two points.

2. Find the unit rate of increase in $ per hour from $20 in 2 hours, to $60 in 4 hours.

3. A person observes a tree from an eye-level of 6ft. The distance from the person's eye to the top of the tree is 28ft, at a $\frac{\pi}{6}$ radians angle with the ground. Find the height of the tree.

4. The "Highseas" ship navigates at a bearing of 29° for 7.5 miles, and then travels $\frac{15\sqrt{3}}{2}$ miles at a bearing of 119°. Find the straight distance from the starting point to the ending point.

(4-20-20-15)

Nine times out of ten the best thing that can happen to a young man is to be tossed overboard and to be compelled to sink or swim,

James A. Garfield

1. The slope of a line is − 1. Find b if the points (5, - 3) and (-2, b) are on the line, then graph the line.

2. A road rises 63 yards for every 300 yards horizontal distance. Find the grade of the road.

3. Find t such that: $10 + t \sin x (\cos x - \sin x) = 10 \sin 2x + 10 \cos 2x$.

4. In the following trigonometric identity determine the value of $(c + 2)$: $(30 \cos^2 x - c) \tan 2x = 30 \sin x \cos x$.

(4-21-20-17)

To measure up to all that is demanded of him, a man must overestimate his capacities.

Johann von Goethe

1. Find the slope of the line: $- 12x + 3y + 8 = 0$.

2. Find the y-intercept of the line of equation: $4x - 3y + 66 = 0$.

3. Use trigonometric identities to find the value of B such that: $B \cos^2 x - 10 \cos 2x = 10$.

4. Find the value of C such that: $(\sqrt{15} \sin x + \sqrt{15} \cos x)^2 = C + 15 \sin 2x$.

(4-22-20-15)

Time is a fixed income and, as with any income, the real problem facing most of us is how to live successfully within our daily allotment.

Margaret B. Johnstone

1. Find the slope-intercept form of the equation of the given line, graph the line and locate the y-intercept: $2x + 3y - 12 = 0$.

2. Find b such that the point $(15, b)$ is on the line that has the slope $m = -\frac{3}{5}$ and the y-intercept equal to 32.

3. Find the value of d if: $5 \cos 3x + 15 \cos x = d \cos^3 x$.

4. Find the value of $(t + 2)$ such that : $20 \sin^3 x = t \sin x - 5 \sin 3x$.

(4-23-20-17)

Reach high, for stars lie hidden in your soul. Dream deep, for every dream precedes the goal.

Pamela Vaull Starr

1. Find the x-intercept of the line passing through the points $(8, -3)$ and $(-4, 6)$.

2. Find the value of b such that the point $(8, b)$ is on the line of equation:
$$y + 4 = 4(x - 1).$$

3. Use trigonometric identities to find d:
$10 \sin 2x (1 - \tan^2 x) = d \cos 2x \cdot \tan x$.

4. Find the value of c, if: $\frac{30 \tan x}{\tan 2x} = c - 15 \tan^2 x$.

(4-24-20-15)

He who would learn to fly one day must first learn to stand and walk and run and climb and dance; one cannot fly into flying.

Friedrich Nietzsche

1. Simplify using properties of exponents: $(-3)^1 + 3^0 - (-3)^2 + 15$.

2. Simplify using properties of exponents, and evaluate the answer for x = 1 and y = -1: $\dfrac{-5^2}{-(3x)^0} \dfrac{yx^3}{y^3}$.

3. Find d by using trigonometric identities if:
   $10 \cos x + d \sin^2 \dfrac{x}{2} = 10$.

4. Find t, if: $\dfrac{15 (\sin x)^2}{1+\cos x} + t \cos x = 15$.

(4-25-20-15)

Nothing is perfect. As we understand imperfections, we are able to love the world and strive for happiness.

Irie Glajar

1. Simplify the expression: $(-3)^2 + (-2)^3 + (3)^1 - (5-5)^2$.

2. Simplify the expression, then evaluate your answer for a = -2 and b = 2: $\dfrac{-13b^2 a^4}{-b^3 (-a)^2}$.

3. Find t if:
   $(4 \sin x + 4 \sin y)(5 \sin x - 5 \sin y) = t - 20 (\sin^2 y + \cos^2 x)$.

4. Use trigonometric identities to simplify:
   $15 \cos x (1 + \tan^2 \dfrac{x}{2}) + 15 \tan^2 \dfrac{x}{2}$.

(4-26-20-15)

They are ill discoverers that think there is no land, when they see nothing but see.

Francis Bacon

1. Subtract $(3x^2 - x + 1)$ from the sum of $(x^2 + 2x)$ and $(4x^2 - x + 1)$. Simplify your answer and evaluate it for $x = -2$.

2. Subtract $(2x^2 - x + 1)$ from the product of $(3x + 2)$ and $(x - 4)$. Simplify your answer and evaluate it for $x = -3$.

3. Evaluate the expression in degrees: $\frac{1}{6} \text{Arccos} \left(-\frac{1}{2}\right)$.

4. Evaluate the expression in degrees: $\frac{\text{Arcsin} (-1)}{18} + 2$.

(4-27-20-17)

Guard well your spare moments. They are like uncut diamonds. Discard them and their value will never be known. Improve them and they will become the brightest gems in a useful life.

Ralph Waldo Emerson

1. A rectangle has the length $(2x + 3)$ and the width $(3x - 2)$. Find the width if the perimeter is 22.

2. The height of a triangle is one unit more than the base. If the height is x^3 find the area of the triangle for $x = 2$.

3. If $\theta = \text{Arctan} \sqrt{3}$, find the value of $\frac{\theta}{3}$ in the first quadrant.

4. With x in the second quadrant and knowing that $x = \text{Sec}^{-1} (-2)$, evaluate $\frac{x°}{8}$.

(4-28-20-15)

The real friend is he or she who can share all our sorrow and double our joys.

B. C. Forbes

1. Multiply and simplify: $(3x - 2)(x - 1)$, then evaluate your answer for x = 2.

2. First simplify the expression, then evaluate it for x = 2:
   $4 + (x + 3)^2$.

3. If $t = \sin\left(\text{Arccos } \frac{\sqrt{3}}{2}\right)$, find the value of 40t.

4. If $p = \cos\left(\text{Arctan } \sqrt{3}\right)$, find the value of 30p.

(4-29-20-15)

Any path is only a path, and there is no affront, to oneself or to others, in dropping it if that is what your heart tells you.

Carlos Castaneda

1. Evaluate the expression for x = 3: $-23 + (2x - 3)^3$.

2. Simplify the expression completely, then evaluate your answer for x = 4:
   $(x^2 + 3x + 9)(x - 3) - 7$.

3. If $k = \tan\left(\text{Arcsin } \frac{\sqrt{3}}{2}\right)$, find $\frac{20k\sqrt{3}}{3}$.

4. Given that $p = \cot\left(\text{Arccos } \frac{1}{2}\right)$, evaluate $(2 + 15p\sqrt{3})$.

(4-30-20-17)

I think and think for months, for years. Ninety-nine times the conclusion is false. The hundredth time I am right.

Albert Einstein

1. Use special products to multiply $(3x - 2)(3x + 2)$, then evaluate the answer for $x = 1$.

2. Use special products to multiply $(3x - 5)^2$, then evaluate the answer for $x = 2$.

3. Evaluate in degrees: $\text{Arccos}\left(\cos \frac{\pi}{9}\right)$.

4. Find the value of the expression in degrees: $2 + \frac{1}{2}\text{Arcsin}\left(\sin \frac{\pi}{6}\right)$.

(5-1-20-17)

On many of the great issues of our time, men have lacked wisdom because they have lacked courage.

William Benton

1. Divide: $\frac{\left(12a^3b - 4a^2b^2 + 20ab\right)}{4ab}$. Then evaluate your answer for $a = -2$ and $b = -6$.

2. Divide: $\frac{\left(3x^2y - 6xy^2 - 3xy\right)}{3xy}$. Then evaluate the answer for $x = 5$ and $y = 1$.

3. Simplify the expression: $-40 \cos\left(\text{Arcsin}\frac{\sqrt{3}}{2} + \text{Arccos}\frac{1}{2}\right)$.

4. Evaluate: $2 - 15\sqrt{3} \tan\left(\text{Arccot}\sqrt{3} - \text{Arctan}\sqrt{3}\right)$.

(5-2-20-17)

Happy the man who early learns the wide chasm that lies between his wishes and his powers.

Johann von Goethe

1. Simplify completely (no negative exponents in the final answer): $-\left(\dfrac{-12x^{-3}y}{3x^{-2}y^3}\right)^{-1}$. Then evaluate the answer for x = 5 and y = - 2.

2. Simplify completely: $-(-5a^2b^{-3})^{(5-5)} + 4$.

3. If $\sin^2 x = \dfrac{1}{4}$ and x is measured in degrees in the first quadrant, evaluate (x – 10).

4. If x, in degrees, is a solution in the interval [0, 360°] for the equation $\cos^2 x - \cos x = 2$, evaluate $\dfrac{x}{12}$.

(5-3-20-15)

One learns by doing the thing; for though you think you know it, you have no certainty until you try.

Sophocles

1. Perform the operation after you write the numbers in proper scientific notation, and write the answer in standard form: $(20 \times 10^{-6})(2.5 \times 10^5)$.

2. Divide and write your answer in scientific notation: $\dfrac{24500000}{1225}$. What is the exponent of 10?

3. Evaluate $\dfrac{x}{3}$, if x is a solution (in degrees) in the first quadrant, with $x \neq 0$, of the equation: $\sin 2x - \sin x = 0$.

4. If x (in degrees) is a solution of the equation: $\cos 2x - \cos x = 0$, such that $0 < x < 180°$, evaluate $\left(\dfrac{x}{8} + 2\right)$.

(5-4-20-17)

Strong people are made by opposition, like kites that go up against the wind.

Frank Harris

1. Factor completely: $6x^3y^2 - 8x^2y - 4xy$. Then evaluate the GCF (greatest common factor) for $x = .5$ and $y = 5$.

2. Factor out the GCF, write it in front of the parentheses, and find the sum of the coefficients in the parentheses: $8a^2b^2 - 6a^2b + 8ab$.

3. Solve the equation: $\sin 2x = \cos 2x + 1$. Then evaluate $\frac{2x}{9}$, with x in the interval $[0, 360°]$ and $\sin x \neq \cos x$.

4. Evaluate $\frac{x}{4}$ if x is the smallest positive angle not equal to 0, in degrees, to satisfy the equation: $\sin 2x + \sin 4x = 0$.

(5-5-20-15)

You will never "find" time for anything. If you want time you must make it.

Charles Baxton

1. Factor completely: $6x^2 - 2xy^2 + 3xy - y^3$. Then evaluate the first factor for $x = 2$ and $y = 1$.

2. Factor completely: $6ab - 3ac + 10b - 5c$. Then evaluate the factor that contains b for $b = 3$ and $c = 0$.

3. Evaluate $(x - 70)$ in degrees for $y = .2$, if $4\sin x - 3 = 5y$, and $x \in [0, 360°]$.

4. If: $\frac{4}{3} \text{Arctan} \frac{x}{2} = \pi$, evaluate $(-\frac{15x}{2} + 2)$.

(5-6-20-17)

What you do not want done to yourself, do not do to others.

Confucius

1. Factor completely: $6x^2 - 13x + 2$. Then evaluate the factor that has the higher coefficient of x, for x = 1.

2. Factor completely: $5m^2 - m + 1$. Then evaluate your answer for m = - 1.

3. If t degrees is the solution in the first quadrant for the equation:
   $4 \cot^2 t - 1 = 3 \cot t$, find $\frac{4}{9}t$.

4. Find the smallest positive value of x in degrees when y = .5, if: 2y = tan 3x.

(5-7-20-15)

Many people take no care of their money till they come nearly to the end of it, and others do just the same with their time.

Johann von Goethe

1. Factor completely: $3t^2 + 8t - 3$. Then evaluate either factor for t = 2.

2. Factor completely: $3n^2 - 10n - 8$. Then find the smallest value of your factors when t = 12.

3. Two angles of a triangle are 45° and 30°. Use the law of sines to find the opposite side of the 30° angle, if the opposite side of the 45° angle is $20\sqrt{2}$.

4. In the triangle from problem #3 find the side opposite to the 45° angle if the side opposite to the 30° angle is $\frac{15\sqrt{2}}{2}$.

(5-8-20-15)

A life spent making mistakes is not only more honorable but more useful than a life spent doing nothing.

George Bernard Shaw

1. Factor completely: $9z^2 - 4$. What is the largest value of your factors for z = 1?

2. Factor completely: $16y^2 - 24y + 9$. What is the value of the simplest factor for y = 3?

3. Two angles of a triangle are 45° and 105°. The height corresponding to the longest side is 10. Find the next longest side of the triangle.

4. Use the information from problem #3 to evaluate the expression: $\frac{3A}{20}(\sqrt{3} - 1)$, where A is the area of the triangle.

(5-9-20-15)

To most of us the real life is the life we do not lead.

Oscar Wilde

1. Factor completely: $1 - 16w^4$. Then evaluate the quadratic factor for w = 1.

2. Factor completely: $4p^2 + 12pq + 9q^2$. Then evaluate the smallest factor for p = 2 and q = 2.

3. Find the area of a triangle if two of its sides are 5 and $\frac{16\sqrt{3}}{3}$, and the angle between these sides is 60°.

4. One angle of a triangle is 30°. One side of this angle is 10 and the area of the triangle is 37.5. Find the other side of the 30° angle.

(5-10-20-15)

The time which we have at our disposal every day is elastic; the passions that we feel expand it, those that we inspire contract it; and habit fills up what remains.

Marcel Proust

1. Factor completely: $52b - 8b^2 - 60$. Then find the smallest possible prime value of your factors for $b = 10$.

2. Factor completely: $18y^2 - 2$. Then find the middle value of the three factors   for $y = 4$.

3. Use the law of cosines to find the opposite side to a $60°$ angle of a triangle, if the adjacent sides of the angle are 15 and $5\dfrac{(3+\sqrt{37})}{2}$.

4. The opposite side of the $30°$ angle of an acute triangle is $5\sqrt{25 - 12\sqrt{3}}$. If one adjacent side of this angle is 20, find the other adjacent side of the angle.

(5-11-20-15)

To find out what one is fitted to do, and to secure an opportunity to do it, is the key to happiness.

John Dewey

1. Solve the equation by factoring: $3x^2 - 13x = 10$. Then identify the positive solution.

2. Solve the equation by factoring: $x(2x - 1) = 6(4x - 2)$. What is the largest solution?

3. Two sides of a triangle are 14 and 16. Find the third side if the area is $15\sqrt{55}$.

4. Two opposite angles of a quadrilateral are each equal to $90°$. If the shortest side of the quadrilateral is $5\sqrt{3}$ and the opposite diagonal to one of the $90°$ angle is $10\sqrt{3}$, find the length of the other diagonal.

(5-12-20-15)

Real life is, to most men ... a perpetual compromise between the ideal and the possible.

Bertrand Russell

1. Solve the equation by factoring: $4x - \frac{5}{x} = 19$. Find the integer . solution.

2. Solve the equation and find the sum of the solutions: $1 + \frac{42}{x^2} = \frac{13}{x}$.

3. If a vector has a 30° inclination from the horizontal, and its horizontal component is $10\sqrt{3}$, find its magnitude.

4. Two vectors make an angle of 60°. Their magnitudes are 14 and $(7 + \sqrt{78})$. Find the magnitude of their difference.

(5-13-20-15)

In order to have great happiness, you have to have great pain and unhappiness – otherwise how would you know when you're happy?

Leslie Caron

1. Solve the equation and identify the positive solution: $\frac{x-1}{10} = \frac{2}{x}$.

2. The product of two consecutive even integers is 224. Find the smallest one of them.

3. Simplify in complex numbers: $\sqrt{-169} + \sqrt{-49}$. What is the multiple of i in your answer?

4. Simplify in complex numbers: $\sqrt{-144} + \sqrt{-16}$ - i. What is the multiple of i in your answer?

(5-14-20-15)

Fear is an instructor of great sagacity, and the herald of all revolutions.
Ralph Waldo Emerson

1. One leg of a right triangle is 7 units longer than the other leg. Find the shorter leg if the hypotenuse is 13.

2. The area of a triangle is 75. Find the base of the triangle if the corresponding height is 5 units shorter than the base.

3. Find the product of the solutions of the equation: $\frac{x}{2}(x-1) = -10$.

4. Solve the equation: $3(x^2 + 21) = 45x$. Then find the sum of the solutions.

(5-15-20-15)

The most important thing about getting somewhere is starting right where we are.
Bruce Barton

1. The length of a pipe that runs diagonally in a rectangular garden is $\sqrt{74}$. If the length of the garden is 2ft shorter than the width, find the length.

2. Two numbers are the first and the fourth multiples of the same positive number. Find the largest multiple of them if the product of the two multiples is 64.

3. Simplify in standard complex form: $(3 - 4i) + (5 - 7i) - \sqrt{-81}$. Then state the absolute value of the coefficient of i.

4. Simplify in standard complex form, a + bi:
$(2 + 5i) - (3 - 6i) - i\sqrt{-256} - i$. Then indicate the real component of your answer, a.

(5-16-20-15)

Treasure the memories of past misfortunes; they constitute our bank of fortitude.

Eric Hoffer

1. Twice the square of a child's age is 29 more than seven times his age two years ago. Find the child's age now.

2. Two positive integers differ by 11. Their product is 42. Find their sum.

3. Simplify in the (a + bi) form: $(5 - 3i)(2 - i)$. Then evaluate the expression: $3a + b + 10$.

4. Simplify in the (a + bi) form: $(2 - 3i)^2$. Then evaluate the expression: $|a + b + 2|$.

(5-17-20-15)

People are always asking about the good old days. I say, why don't you say the good "now" days? Isn't "now" the only time you're living?

Robert M. Young

1. Find the positive integer for which the expression is undefined:
   $$\frac{(x-3)}{(x^2 + 3x - 40)}.$$

2. Simplify the expression and then evaluate the answer for x = - 4:
   $$\frac{(3x-2)}{(3x^2 + 7x - 6)} + 19.$$

3. Simplify and write your answer in the (a + bi) form: $5\left[\frac{5-2i}{2+i}\right]$.
   Then evaluate the expression: $a + |b| + 3$.

4. Simplify in the form (a + bi): $\left(\frac{4+i}{i}\right)^2$. Then find $(|a| + 2)$.

(5-18-20-17)

87

Happiness belongs to those who are sufficient unto themselves. For all external sources of happiness and pleasure are, by their very nature, highly uncertain, precarious, ephemeral and subject to chance.
Arthur Schopenhauer

1. For what value of y is the expression: $\left(\frac{y-5}{y^2-10y+25}\right)$ undefined?

2. Simplify: $\frac{(1-b^2)}{(3b^2-b-2)}$. Then evaluate your answer for b = $-\frac{39}{58}$.

3. Simplify: $-10(i^{19}-i^{25})$. Then find the coefficient of i.

4. Simplify: $2 - 15i^{-30}$

(5-19-20-17)

Acceptance is not submission; it is acknowledgment of the facts of a situation. Then deciding what you're going to do about it.
Kathleen Casey Theisen

1. Simplify the expression: $\frac{(x^2-y^2)}{(x^2-xy-2y^2)} \cdot \frac{(2y-x)}{(x-5y)}$. Find the value of x for which the answer is undefined when y = 1.

2. Simplify: $\frac{(x-2)}{(x^3-xy^2)} \div \frac{(x^2-4)}{(3x-3y)}$. Then evaluate half of the denominator of the answer for x = 2 and y = 3.

3. Simplify: $-5(\cos 30° + i \sin 30°) \cdot 4(\cos 60° + i \sin 60°) \cdot i$.

4. Simplify: $10\sqrt{3} \cdot \frac{cis\ 60°}{cis\ 30°}$. Write your answer in (a + bi) form and indicate the real component (a) of your answer.

(5-20-20-15)

The U.S. constitution doesn't guarantee happiness, only the pursuit of it. You have to catch up with it yourself.

Benjamin Franklin

1. Simplify the expression, then evaluate the answer for $x = \frac{11}{5}$:

$$\frac{x}{(x^2 - 4x+4)} + \frac{-2}{(x^2 - 4x+4)}.$$

2. Perform the operation, simplify your answer and evaluate it for $x = -\frac{104}{21}$:

$$\frac{2x-3}{(x^2 - 25)} - \frac{x+2}{(x^2 - 25)}.$$

3. Simplify $(2 - i)^3$ to the form (a + bi). Then evaluate: $|2b + a|$

4. Use Moivre's theorem to solve the equation for x:
$(2 + 2i)^4 x + 960 = 0.$

(5-21-20-15)

Problems are the cutting edge that distinguishes between success and failure. Problems ... create our courage and wisdom.

M. Scott Peck

1. Perform the operations, simplify your answer, then find the value of x that makes the numerator equal to 0: $\frac{5}{x-2} + \frac{x-40}{(x^2 - 4)}.$

2. Subtract the fractions, simplify the answer, then find the absolute value of the numerator for $x = 4$: $\frac{-2}{3x-6} - \frac{x}{(x^2 - x - 2)}.$

3. Two different vectors representing real numbers in the complex plane have each a magnitude of 10. Find the distance between their endpoints.

4. Find the magnitude of the vector corresponding to the graph of the complex number: $3\sqrt{5} + 6i\sqrt{5}.$

(5-22-20-15)

If you have to support yourself, you had bloody well better find some way that is going to be interesting.
Katharine Hepburn

1. Perform the operations, simplify your answer, and find the value of x for which the *answer* is undefined:
$$\frac{2}{3x-15} - \frac{x+2}{(x^2 - 3x - 10)}.$$

2. Simplify the rational expression, then evaluate it for
x = - 1: $\quad \frac{1}{x} - \frac{3}{x^2} - \frac{27}{x^3}.$

3. A complex number in trigonometric form is:
$25[\cos (\text{Arctan}\frac{4}{3}) + i \sin (\text{Arctan}\frac{4}{3})]$. Find the length of its vertical component.

4. A vector of magnitude $10\sqrt{3}$ forms an angle of 60° with the vertical axis. Find the length of its horizontal component.

(5-23-20-15)
Real generosity toward the future lies in giving all to the present.
Albert Camus

1. Simplify the complex fraction, then evaluate the denominator of your answer for t = 1: $\quad \dfrac{\left(\frac{1}{t} - \frac{1}{t^2}\right)}{\left(\frac{2}{t} + \frac{3}{t^3}\right)}.$

2. Simplify: $\quad \dfrac{\left(\frac{3m-2}{2} + 1\right)}{\frac{m}{16}}.$

3. If a complex number in trigonometric form is: 40 cis 60°, find 'a' in its (a + bi) form.

4. Find b from the (a + bi) form, if the trigonometric form of a complex number is: $10\sqrt{3}$ cis 420°.

(5-24-20-15)

The last, if not the greatest, of the human freedoms: to choose their own attitude in any given circumstance.

Bruno Bettelheim

1. Solve for y: $\dfrac{2}{y-1} - \dfrac{y-1}{(y^2-1)} = \dfrac{2}{y+1}$.

2. Find the absolute value of the product of the solutions of the equation:
$$\frac{x}{x-3} + \frac{3}{x+3} = \frac{16}{(x^2-9)}.$$

3. Find the absolute value of the real solution of the equation: $t^3 + 8{,}000 = 0$.

4. Find the product of the x-intercepts of the graph of the equation represented in the following polar form:
$$\frac{r}{(r\cos\theta-3)} = \frac{(r\cos\theta-5)}{\sin\theta}.$$

(5-25-20-15)

I believe that a simple and unassuming manner of life is best for everyone, best both for the body and the mind.

Albert Einstein

1. In the following rational equation, find the value of c if x = 7:
$$\frac{1}{c} + \frac{1}{x-2} = \frac{2}{2x-9}.$$

2. Ian alone can clean the family kitchen in 39 minutes. Together with his brother they can clean the kitchen in 15.6 minutes. How long would his brother take to clean the kitchen by himself?

3. Find the four fourth roots of 64. Then find $\frac{1}{9}$ of the second largest angle in the trigonometric representations of your answers.

4. Solve the equation: $x^2 = -i$. Then find $\frac{1}{9}$ of the smallest angle in the trigonometric representations of your answers.

(5-26-20-15)

The secret of being miserable is to have leisure to bother about whether you are happy or not. The cure for it is occupation.

George Bernard Shaw

1. While Eric drives 200 miles at 50 mph, John drives 220 miles. How much faster does John drive?

2. Two printers can print a manuscript together in 18 minutes. One of them can do the job alone in 54 minutes. How long will it take the other computer to print the manuscript alone?

3. Find the equivalent equation in rectangular coordinates if the polar form is: $r^2 = r \cos \theta$ ($r \cos \theta - 1$). Then find the absolute value of x if $y = 2\sqrt{5}$.

4. In problem #3 find the positive value of y if $x + 225 = 0$.

(5-27-20-15)

We have too many sounding words and too few actions that correspond with them.

Abigail Adams

1. If $y = 1$, find the value of b in the system of equations:
$3x - y = b$
$x + y = 3$

2. Find the sum of x and y from the system:
$x - y = -4$
$3x - 2y = 4$

3. Simplify: $\dfrac{\left(2^7 \cdot 5^{-3}\right)}{\left(2^5 \cdot 5^{-4}\right)}$.

4. Simplify: $\dfrac{\left(9^{\frac{3}{2}} \cdot 5^{\frac{1}{2}}\right)}{\left(3^2 \cdot 5^{-\frac{1}{2}}\right)} + 2$

5-28-20-17)

Have patience with all things, but chiefly have patience with yourself. Do not lose courage in considering your own imperfections, but instantly set about remedying them – every day begin the task anew.

Saint Francis de Sales

1. Use substitution to solve the system:

$$4a + b = -7 \qquad 2a + 3b = 9.$$

What is the value of b?

2. Find the value of t, such that the following system is inconsistent:

$$3x - y = t$$
$$-x + \frac{1}{3}y = -9\frac{2}{3}$$

3. Graph the function: $y = 2^{x-3}$. Then find the product of x and y when y = 4.

4. Point (x, 9) is on the graph of the function:
$y = 3^{-\frac{x}{3}}$. Evaluate: $(|x| + y + 2)$.

(5-29-20-17)

Once a decision was made, I did not worry about it afterward.

Harry S. Truman

1. Use elimination to solve the system for m:
$2m - 3n = 22.$
$3m + 2n = -23$

2. Solve the system by any method, then find the sum of x and y: $y - x = 2$
$3x - 2y = 10$

3. Graph the function: $y = \log_2(x - 4)$, and find x if y = 4.

4. If the point (t, 5) is on the graph of the function: $y = \log_3 x + 4$, evaluate $(5t + 2)$.

(5-30-20-17)

Getting an idea should be like sitting down on a pin; it should make you jump up and do something.

E. L. Simpson

1. The average of two integers is equal to -1. Three times the second is one unit larger than two times the negative of the first. Find the second integer.

2. Two angles are complementary. One angle is 28° smaller than the other. Find the smaller angle.

3. Convert the equation to logarithmic form and evaluate the logarithm: $2^x = 32$. Then find the value of (4x).

4. Convert the equation to logarithmic form: $10^{-x} = 0.001$. Then evaluate (2 + 5x).

(5-31-20-17)

With begging and scrambling we find very little, but with being true to ourselves we find a great deal more.

Rabindranath Tagore

1. The difference of two numbers is 5. Their average plus $\frac{1}{2}$ equals 4. Find the larger number.

2. Find the smaller of the two numbers from problem #1.

3. Write in exponential form and find b: $\log_b \frac{125}{64} = -3$. Then, evaluate (25b).

4. Given the equation: $\log_{\frac{1}{3}} 243 = y$, evaluate (2 - 3y).

(6-1-20-17)

Knowing is not enough, we must apply. Willing is not enough, we must do.

Johann von Goethe

1. There are four fewer dimes than nickels in a box. If there are only dimes and nickels, and they are worth a total of $0.5, how many nickels are there?

2. How many dimes are there in the box from problem #1?

3. Evaluate: $\log_2 32 + 15\log_{23} 23 - 25\log_4 1$.

4. Evaluate: $\dfrac{-\log_3 27}{\log_{\frac{1}{2}} 32} + 2$.

(6-2-20-17)

Do each daily task the best we can; act as though the eye of opportunity were always upon us.

William Feather

1. How many pounds of Mexican coffee worth $9.75 per pound should be mixed with Brazilian coffee worth $10.25 per pound, to make 9 pounds of a blend worth $9.92 per pound? (Round to the nearest pound.)

2. How many pounds of the Brazilian coffee should be used in problem #1?

3. Evaluate: $(\log 100)^2(\log 100 + \log 1000)$.

4. Evaluate: $2 + (\ln e^3)(\ln e)^3(\ln e^5)$.

(6-3-20-17)

Men's real life is happy, chiefly because he is ever expecting that it soon will be so.

Edgar Allan Poe

1. There are two children in a family. Ten years from now the sum of their ages will be 30. Two years ago, the difference of their ages then was 2. How old is the older child?

2. How old is the younger child in problem #1?

3. Simplify using properties of logarithms, then evaluate the answer for $x = y = 4$: $2(\frac{2}{3}\log_2 x^3 + \frac{3}{2}\log_2 y^2)$.

4. Simplify using properties of logarithms, then evaluate the answer for $x = e^2$ and $y = e^3$:
$2 + 5(\ln x^6 - \ln y^3) - 3\ln x + 2\ln y$.

(6-4-20-17)

We succeed in enterprises which demand the positive qualities we possess, but we excel in those which can also make use of our defects.

Alexis de Tocqueville

1. Multiply and simplify the final answer: $(-\sqrt{4})(-\sqrt{9})$.

2. Simplify the radical and indicate the integer coefficient: $\sqrt{125x^2}$.

3. Solve the exponential equation, then evaluate (10x): $4^{2x} = 256$.

4. Solve the equation and evaluate (3x + 2): $3^{x-2} - 27 = 0$.

(6-5-20-17)

What fate can be worse than to know we have no one but ourselves to blame for our misfortunes?

Sophocles

1. Simplify the expression: $-(-\sqrt{36})$.

2. Multiply the radicals and simplify: $\sqrt{3x}\ \sqrt{12x}\ \sqrt{1}$. What is the coefficient of x?

3. Solve the equation, then evaluate (5x): $3^x - 81 = 0$.

4. Solve the exponential equation: $5^{x-1} = 625$. Then evaluate $(2 + 3x)$.

(6-6-20-17)

Man does not simply exist, but always decides what his existence will be, what he will become in the next moment.

Viktor Frankel

1. Simplify the expression and indicate the integer in front of the radical: $\sqrt{108t}$.

2. Multiply and simplify: $\sqrt{14}\ \sqrt{21}$. What is the integer part of your answer?

3. Solve the equation: $4^{x+10} = 8^x$.

4. Solve the equation: $3^{-x+45} = 9^x$.

(6-7-20-15)

Simple style is like white light. It is complex, but its complexity is not obvious.

Anatole France

1. Simplify the radical: $\sqrt{(64x^{12}y)}$. What is the exponent of x?

2. Multiply and simplify: $\sqrt{2x^2}\,\sqrt{8x}\,\sqrt{4x}$. Find the coefficient of x in the final answer.

3. Solve the equation: $\log_2(x-4) = \log_2(x+12)$ - 1.

4. Solve the logarithmic equation: $1 - \log(2x+120) = -\log x$.

(6-8-20-15)

As long as I have a want, I have a reason for living. Satisfaction is death.

George Bernard Shaw

1. Multiply and simplify: $\sqrt{3a^2}\,\sqrt{(5a^4b)}\,\sqrt{(30a^6d)}$. What is the exponent of a?

2. Multiply and simplify: $\sqrt{3t}\,\sqrt{12tz}$. Evaluate your answer for t = 1.5 and z = 1.

3. Solve the logarithmic equation for t: $\ln(3t) - \ln(t-5) = \ln 4$.

4. Solve for z: $\log_3 z = 3 - \log_9 z$. Then evaluate $(1\frac{2}{3}z + 2)$.

(6-9-20-17)

Many of life's failures are men who did not realize how close they were to success when they gave up.

Thomas A. Edison

1. Solve by the principle of square roots. Identify the positive solution: $72 - 2z^2 = 0$.

2. Solve by the principle of square roots. Then find the sum of the absolute values of the solutions: $-3y^2 + 75 = 0$.

3. Solve the logarithmic equation: $\log x + \log (x - 15) = 2$.

4. Solve for m, then evaluate $(3m + 2)$:
   $\ln (m - 2) - \ln 3 = \ln 5 - \ln m$.

(6-10-20-17)

When I feel difficulty coming on, I switch to another book I'm writing. When I get back to the problem, my unconscious has solved it.

Isaac Asimov

1. Solve by the principle of square roots: $2t^2 - 4t + 2 = 50$. Then identify the positive solution.

2. Use the principle of square roots to solve the equation, then state the larger solution: $n^2 - 10n + 25 = 36$

3. Solve the equation: $\log_7(2x - 37) = -\log_7 x + \log_7 60$.

4. Solve both equations, then evaluate $(mn + 2)$:
   $\ln (2m - 5) = 0$, $2^{3n - 8} = 128$.

(6-11-20-17)

Write it on your heart that every day is the best day in the year.
Ralph Waldo Emerson

1. Solve by the quadratic formula: $2x^2 - 9x - 18 = 0$. Indicate the positive solution.

2. Solve the equation by any method and find the sum of the solutions: $(x - 13)(x + 1) = -40$.

3. Solve the logarithmic equation for b, then evaluate (4b): $\log_b 625 - 4 = 0$.

4. Solve the equation for x: $\log_x 3 = 1 - \log_x 5$.

(6-12-20-15)

When one door closes another opens. But we often look so long and so regretfully upon the closed door that we fail to see the one that has opened for us.
Alexander Graham Bell

1. Solve the equation by the quadratic formula and find the absolute value of the smaller solution: $3t^2 + 16t = 12$.

2. Solve the equation, then evaluate (26z): $z(z - 1) = -\frac{1}{4}$.

3. Find f(9) if: $f(x) = 5 \log_3 (x + 72)$.

4. Solve the logarithmic equation for p: $\ln p - \ln (p - 10) = \ln 3$.

(6-13-20-15)

Acceptance of what has happened is the first step to overcoming the consequences of any misfortune.

William James

1. Solve the equation and identify the larger solution: $1 - \frac{4}{x} = \frac{12}{x^2}$.

2. One leg of a right triangle is $\sqrt{29}$. The other leg is one unit shorter than the hypotenuse. Find the longer leg.

3. If: $f(x) = \log_2(x - 4) - \log_4(x + 44)$, find the value of x such that $f(x) = 1$.

4. Solve for x: $2 + \log_5 1 = \log(x - 5) + \ln e$.

(6-14-20-15)

Know what you want to do, hold the thought firmly, and do every day what should be done, and every sunset will see you that much nearer the goal.

Elbert Hubbard

1. The two bases of a trapezoid are 5 and 10. Find the height of the trapezoid if its area is 45.

2. The diagonal of a rectangular garden is 20ft and its width is $5\sqrt{7}$ft. Find the length of the garden.

3. A point P(x,y) is on the graph of the logarithmic function: $f(x) = 5 \log_2(x - 2)$. Evaluate (x·y) when x = 4.

4. Evaluate the function: $f(x) = \ln x^4 + \ln x^3 + 4 \ln x^2$ when x = e.

(6-15-20-15)

What we do upon some great occasion will probably depend on what we already are, and what we are will be the result of previous years of self-discipline.

P. Liddon

1. Two positive numbers are different by 10. Find the smaller number, if its square is 20 larger than the other number.

2. Find the larger of the two numbers from problem #1.

3. Solve for x: $16 \cdot 2^x = 4^{x-12}$. Then, find $(x - 8)$.

4. Solve the logarithmic equation for p when x = 18:
$$3 - \log_p(8x + 81) = \log_p(x - 3).$$

(6-16-20-15)

Events, circumstances, etc. have their origin in ourselves. They spring from seeds which we have sown.

Henry David Thoreau

1. Twice a positive number, minus 30 times its reciprocal, is equal to 7. Find the number.

2. Evaluate $(2t + 5)$ if t is the number found in problem #1.

3. Find the integer value of x in the interval [2, 8] that is a solution of the equation: $\log_2 x = 2^{3x-11}$. Then evaluate $(5x)$.

4. Solve the equation, then evaluate $(2 + 5x)$: $\log_3 x - 3 = -\log_9 x^4$.

(6-17-20-17)

Intermediate Algebra (1, 2) and Precalculus (3, 4)

Once you say you're going to settle for second, that's what happens to you.

John F. Kennedy

1. How many rational numbers are there in the set
$\{-3, \pi, -2.3, \frac{1}{3}, e, 0, 2\frac{2}{3}, 1, \sqrt{3}\}$.

2. Find the simplest form of the only rational number from the following set:
$$\{-\sqrt{2}, -\pi, \sqrt{8}, -2\sqrt{5}, 6\sqrt{9}\}.$$

3. Find the positive x-intercept for the graph of the equation:
$x^2 + 4y^2 = 400$.

4. Find the y-intercept larger than zero for the graph of:
$-9x^2 + y^2 = 225$.

(6-18-20-15)

Adversity leads us to think properly of our state, and so is most beneficial to us.

Samuel Johnson

1. Simplify: $-\frac{10}{3} \div \left(\frac{5}{-9}\right)$.

2. Simplify the absolute value expression: $\left| -29 + \left(\frac{7}{3} - \frac{1}{3}\right) + 8 \right|$.

3. Find the positive x- and the y-intercepts of the equation $x^2 - 5x + y^2 - 4y = 0$, and then determine their product.

4. Find the y-intercept of the function: $y = (2x - 3)^2 + 6$.

(6-19-20-15)

Decision is a sharp knife that cuts clean and straight; indecision, a dull one that hacks and tears and leaves ragged edges behind it.

Gordon Graham

1. Simplify the complex fraction: $\dfrac{\frac{12}{5}}{\frac{6}{15}}$.

2. Simplify the expression: $13 + \left| -\frac{8}{5} + \frac{3}{5} \right| + 2|-3|$.

3. Find the slope-intercept form of the line, graph the line, and indicate its   y-intercept: $3y + 2x = 60$.

4. The point (- 4, 9) is on a line that is perpendicular to the line from problem #3. Find the y-intercept of the new line.

(6-20-20-15)

Happiness is not a state to arrive at, but a manner of traveling.

Margaret Lee Runback

1. Simplify the complex expression: $\dfrac{\left(\frac{3}{4} - \frac{1}{2}\right)}{\left(\frac{1}{6} - \frac{1}{8}\right)}$.

2. Simplify: $\dfrac{3 + \left(8 - 2^2\right)}{\left(3^2 - 4\,(2)\right)} \cdot \dfrac{\left(5 - 5^2 - 3^0\right)}{\left(-3^2 + 2\right)}$.

3. The point (8, b) is on the graph of a line that is parallel to the line $y = 3x$. If the new line has the y-intercept (- 4), find the value of b.

4. The x-intercept of a line is 30 and the line passes through the point (2, 14), Find the equation of the line, graph it, and indicate its y-intercept.

(6-21-20-15)

Happiness is when what you think, what you say, and what you do are in harmony.

Mahatma Gandhi

1. Simplify: $\dfrac{(37-2^5)}{(2^3-7)} \div \left(\dfrac{1}{3}+\dfrac{1}{2}\right).$

2. Simplify: $\dfrac{(3^2+2)}{(4^2-3(5))} \div \left(\dfrac{3}{5}-\dfrac{1}{10}\right).$

3. Find the y-intercept of the line with the slope (.8), passing through   (- 5, 16).

4. Find the x-intercept of a line containing the point (10, 2) and having the slope (- .4).

(6-22-20-15)

Courage is what preserves our liberty, safety, life, and our homes and parents,our country and children. Courage comprises all things.

Plautus

1. Simplify: $\dfrac{(3^2-2)}{2(3)-4} - \dfrac{(5^2-4(5))}{(-5^2+23)}.$

2. Simplify: $\dfrac{(-7-(-3)^3)}{9-2^3} + \dfrac{15-3^1}{5-3^0}.$

3. Graph the line and find its y-intercept: $\dfrac{1}{4}x - \dfrac{2}{5}y = -8.$

4. A line is perpendicular to y = - 5x + 10. Find its x-intercept if its y-intercept is –3.

(6-23-20-15)

But are not this struggle and even the mistakes one may make better, and do they not develop us more, than if we kept systematically away from emotions?

Vincent van Gogh

1. Simplify the expression: $-3^2 + [4^2 - 3^2 - 2^2]$. Then find the absolute value of your answer.

2. Simplify: $(9^2 - 3^2)(-3^2 + 4^2 - 1)^0 \div (2^2 - 1)$.

3. Find the distance between the points $(17, 2\sqrt{7})$ and $(2, 7\sqrt{7})$.

4. Find the positive value of b if the distance between the points $(-2, b)$ and $(2, -1)$ is $4\sqrt{17}$.

(6-24-20-15)

Endurance is one of the most difficult disciplines, but it is to the one who endures that the final victory comes.

Buddha

1. Simplify: $4^3 - (2(5) - 2^3)(5^2 + (-2)^2)$.

2. Simplify: $-2^2 (6(5) - 5) \div (-5 + 3)^2$, then find its opposite.

3. Find the radius of the circle: $x^2 + y^2 = 30x + 10y\sqrt{7}$.

4. Find the product of the x- and y-coordinates of the center of the circle:
$x^2 + y^2 = 6x + 10y - 25$.

(6-25-20-15)

Nothing can bring you peace but yourself.
Ralph Waldo Emerson

1. Simplify the expression, then evaluate the answer for
   $x = 4\frac{4}{7}$: $\frac{1}{3}(2x - 1) + \frac{1}{2}(x + 2)$.

2. Simplify the expression, then evaluate your answer for
   $x = 217$: $\frac{3}{5}(x + 3) - \frac{1}{2}(x - 5)$.

3. A circle of radius $10\sqrt{2}$ is tangent to both the x- and the y-axes. Find the distance between its x- and y-intercepts.

4. The circle: $x^2 - 4x + (y - b)^2 = 221$ is tangent to the x-axis. Evaluate $(2 + b)$.

(6-26-20-17)

The first thing to learn in intercourse with others is non-interference with their own peculiar ways of being happy, provided those ways do not assume to interfere with ours.
William James

1. Simplify the expression, then evaluate the answer for
   $x = 15\frac{3}{4}$: $\frac{1}{2}(\frac{3}{5}x - \frac{1}{2}) + \frac{1}{5}(\frac{1}{2}x - \frac{1}{4})$.

2. Simplify the expression in fraction form: $\frac{3}{5}(\frac{2}{5}x - \frac{1}{2}) - (\frac{1}{5}x + 1)$.
   Then evaluate the numerator of the answer for $x = 46$.

3. Write an equation of the circle with the center at (- 6, 8), passing through the origin. Then find its diameter.

4. A circle is passing through the points (0, 15) and (15, 0), and it is tangent to the x- and y-axis. Write the equation of the circle and indicate its radius.

(6-27-20-15)

There is in the worst of fortune the best of chances for a happy change.

Euripides

1. Find the value of the expression for t = 2: $-\dfrac{(14-2t^2)}{t-3}$.

2. Evaluate the expression for z = 3: $\dfrac{(3z^2 - z + 4)}{z-2}$.

3. A circle contains the points (6, 18), (16, 8), and passes through the origin. Find the equation of the circle and determine its diameter.

4. The endpoints of a diameter of a circle are $(10, 20\sqrt{2})$ and the origin. Find an equation of the circle, and determine its radius.

(6-28-20-15)

Each citizen should play his part in the community according to his individual gifts.

Plato

1. Evaluate the expression for y = - 3: $\dfrac{(-y^3 + 4y + 3)}{2y+9}$.

2. Solve the equation: $\dfrac{m+1}{5} + \dfrac{m-19}{10} = \dfrac{m-5}{3}$. Then find (30 + m).

3. D is the diameter of the circle with the center at (- 1, 1), containing the point (- 3, 5). Evaluate $D\sqrt{5}$.

4. A circle of center (a, b) is circumscribed to the triangle of vertices (0, 5), (3, 0), and (0, 0). Find an equation of the circle, then evaluate (4ab + 2).

(6-29-20-17)

All the great things are simple, and many can be expressed in a single word: freedom; justice; honor; duty; mercy; hope.

Sir Winston Churchill

1. Solve the equation; $\frac{x}{3} + \frac{x-1}{5} = \frac{x}{2}$.

2. Solve for n: $\frac{2x-5}{5} - \frac{x+5}{7} = 6$.

3. The points (- 5, b), (2, - 1), and (- 2, 11) are on the graph of the linear function f(x). Find the equation of f and determine the value of b.

4. The slope of a linear function g(t) is (- 2). Given that g(8) = - 1, write the equation of g, graph the function, and determine its y-intercept.

(6-30-20-15)

No trumpets sound when the important decisions of our life are made. Destiny is made known silently.

Agnes de Mille

1. Three times a number, minus 7, is equal to twice the number. Find the number.

2. A number plus a third of the number, minus $\frac{4}{3}$ is equal to 0. Find the number.

3. Given that $f(x) = \frac{4x+B}{-x+3}$, find B if f(5) = - 20.

4. Find (2 + A) if: $g(z) = \frac{3A}{z^2} - \frac{1}{z} - \frac{4}{3}$ and g(- 3) = 4.

(7-1-20-17)

We should draw from the heart of suffering itself
the means of inspiration and survival.

Sir Winston Churchill

1. A number subtracted from three times that number, plus 5, equals 12 more than the number. Find the number.

2. Adam and Drew win prizes at a math competition. Their total is $375. If 8 times the amount Drew wins, plus four times the amount Adam wins would be $2,000, how many times is Adam's amount larger than Drew's?

3. The point (6, b) is on the graph of $f(x) = \frac{4x}{x-5} - 4$. Find the value of b.

4. If $g(t) = \frac{t-1}{t+2} + 13$, find $(g(-5) + 2)$.

(7-2-20-17)

There are two ways of meeting difficulties.
You alter the difficulties or you alter yourself to meet them.

Phillis Bottome

1. Cici has three fewer dimes than quarters. If the total amount she has in dimes and quarters is $1.45, how many coins does she have over all?

2. If $C = 39\frac{7}{9}$ and $C = \frac{5F-32}{9}$, evaluate $\frac{F}{26}$.

3. Given that $f(x) = \frac{3+x}{x+1}$, find $f(x - 2)$, then evaluate your answer for $x = \frac{21}{19}$.

4. Let $g(x) = \frac{-x}{x+2}$. Evaluate $(- g(-\frac{15}{7}) + 2)$.

(7-3-20-17)

We should consider every day lost on which we have not danced at least once.

Friedrich Nietzsche

1. The perimeter of a rectangle is 22. If the length is 3 units longer than the width, find the length.

2. Find the width of the rectangle from problem #1.

3. If $h(t) = \frac{t^2}{t+1}$, evaluate $h(10 + 2\sqrt{30})$.

4. Given that $k(z) = -z^2 + z$, find the positive value of z such that $k(z) = -20$, then evaluate $(2 + 3z)$.

(7-4-20-17)

Too many people miss the silver lining because they're expecting gold.

Maurice Setter

1. Find the smaller of two complementary angles, if the larger angle is 13° more than ten times the smaller.

2. If 60° plus three times the smaller of two supplementary angles is equal to the larger angle, how many times is the larger angle larger than the smaller?

3. Given $f(x) = \sqrt{x^2 - 9}$ and $x - f(x) = 1$, find the product of x and f(x).

4. If $f(x) = 2\sqrt{x^2 + 21}$, find $x > 0$ such that $f(x) = 22$, then evaluate $(1.5x + 2)$.

(7-5-20-17)

Do you know what the greatest test is? Do you still get excited about what you do when you get up in the morning?

David Halberstam

1. Find the lowest value of x if: $\frac{1}{2}(x-1) + \frac{1}{3}(x+2) \geq 6$.

2. Solve the inequality, graph the solution on a number line, and indicate the largest possible value of x: $\frac{2}{3}(x+3) - \frac{1}{6}x \leq 5$.

3. Find $\sqrt{g(25)}$ if g(x) = $|x^2 - 225|$.

4. If f(t) = + $|400 - t^2|$, evaluate $(2 + \sqrt{f(25)})$

(7-6-20-17)

Fair play with others is primarily not blaming them for anything that is wrong with us.

Eric Hoffer

1. Find the smallest integer number that is a solution of the inequality: $\frac{2x-3}{10} > 1$.

2. What is the upper limit of the solution interval of the inequality: $\frac{5-3x}{8} \geq -2$?

3. Given that f(x) = $\frac{x^2}{x^2-6} + \frac{92}{5}$, evaluate f(- 4).

4. If g(t) = $\frac{3t-1}{2t+1}$, find g(2t), then evaluate (2 + 15·g(2t)) for t = 1.

(7-7-20-17)

The only ones among you who will be really happy are those who will have sought and found how to serve.

Albert Schweitzer

1. Find the midpoint of the solution interval for the double inequality: $-\frac{2}{5} < \frac{2}{5}(x-5) < 2.$

2. Solve the double inequality, graph the solution on a number line, and find the point in your solution set that is $\frac{1}{4}$ of the way to the right-side limit:
$$-1 \le \frac{x-9}{3} \le \frac{5}{3}.$$

3. If [a. b] is the domain of the function $f(x) = \sqrt{-x^2 + 9x - 20}$, find the value of (a·b).

4. What is the right-side limit of the domain of the function: $g(z) = -\sqrt{60 - 4z}.$

(7-8-20-15)

You cannot step twice into the same river, for other waters are continually flowing on.

Heraclitus

1. Graph the equation $y = 2|x| + 7$, then indicate the lowest value of y.

2. Graph the equation $y = 3 - |x|$ and evaluate the absolute value of the product of its x-intercepts.

3. For what value of x is the function $f(x) = \frac{-\sqrt{x}}{|x|}$ equal to $\left(-\frac{\sqrt{5}}{10}\right)$.

4. Given $g(x) = \frac{3|x|}{-x} + 18$, evaluate $(g(1001) + 2)$

(7-9-20-17)

Nature magically suits a man to his fortunes, by making them the fruit of his character.

Ralph Waldo Emerson

1. Graph the equation $x = y^2 - 3$ by plotting points, then locate the point on the x-axis that is 10 units to the right side of the lowest value of x.

2. Graph the equation $y = x^2 - 25$. Find the absolute value of each of its x-intercepts, then calculate their sum.

3. If $f(t) = \sqrt{(t^2 + 4)} + 18$, find the lowest value in the range of f (t).

4. Find the smallest positive integer value excluded from the domain of the function: $g(x) = \frac{-x}{\sqrt{(x^2 - 13x - 30)}}$.

(7-10-20-15)

The prizes go to those who meet emergencies successfully. And the way to meet emergencies is to do each daily task the best we can.

William Feather

1. Find the y-intercept of the line with the slope $\frac{2}{3}$ and passing through $(-10, \frac{1}{3})$.

2. The point (5, b) is on the line with y-intercept 17 and passing through the points of coordinates (2, - 5) and (- 3, 1). Find the value of b.

3. For the piece function: $f(x) = \begin{cases} -3x - 4 \text{ if } x < 0 \\ x + 2 \text{ if } x \geq 0 \end{cases}$, find the value of f(- 8).

4. For the function in problem #3, evaluate (f(4) + 2).

(7-11-20-17)

Courage is being scared to death ... and saddling up anyway.
John Wayne

1. A line is perpendicular to the line of equation $2y - 3x = 5$, and passes through the point $(3,5)$. Find the y-intercept of the new line.

2. If point $(t, -1)$ is on the new line from problem #1, find the value of t.

3. Let m be the average rate of change from 3 to 7 for the function $f(x) = x^2$. Evaluate $(2m)$.

4. Find the average rate of change from 3 to x for the function $g(x) = x^2 - 3x$. Then evaluate it for $x = 15$.

(7-12-20-15)

The secret of being miserable is to have leisure to bother about whether you are happy or not.
George Bernard Shaw

1. A line is parallel to the line of equation $3x - 4y = 1$, and passes through the point $(9, \frac{3}{2})$. Find the x-intercept of the new line.

2. The point $(24\frac{1}{3}, b)$ is on the graph of the new line from problem #1. Find b.

3. If n is the average rate of change from 2 to x for the function $f(x) = 8x - 4x^2$, evaluate n for $x = -5$.

4. Let ARC be the simplest form of the average rate of change from 3 to p for the function $h(p) = p^3$. Evaluate $(ARC - 4)$ for $p = -5$.

(7-13-20-15)

Most men pursue pleasure with such breathless haste that they hurry past it.

Soren Kierkegaard

1. Graph the inequality in two variables: $\frac{1}{14}x + \frac{1}{7}y \leq 1$. Then find the largest value of y such that the point (x, y) that satisfies the inequality, has x $\geq$ 0.

2. In problem #1, find the largest value of x such that the point (x, y) has y $\geq$ 0.

3. Determine if $f(x) = x^2 - 5$ is odd, even, or neither, then evaluate f(- 5).

4. Determine if $g(t) = t^3 - 4t$ is odd, even, or neither. Then find b if the point (3, b) is on the graph of the function.

(7-14-20-15)

The confidence which we have in ourselves gives birth to much of that which we have in others.

Francois de La Rochefoucauld

1. Graph the inequality $\frac{1}{14}x - \frac{5}{2}y \geq \frac{1}{2}$. If the point (t, 0) satisfies the inequality, what is the smallest integer value of t?

2. Graph the inequality $\frac{1}{4}x + \frac{1}{6}y < 1$. Then evaluate (a·b) if the points (a, 0) and (0, b) have the largest integer coordinates in the solution of the inequality.

3. Show that $f(z) = 3z^4 - 7z^2$ is even, then evaluate f(2) and f(- 2).

4. Determine if the function $f(x) = \sqrt{x^2 + 81}$ is odd, even, or neither. Then find c if (12, c) is a point on the graph of f(x).

(7-15-20-15)

Help yourself and heaven will help you.

Jean de La Fontaine

1. How many elements does the domain of the following relation have?

   R = {(3, -1), (0, 2), (1, 3), (4, -5), (- 1, 3), (- 3, 1), (- 5, 4), (1, -2)}.

2. If n is the number of elements in the range of the following relation, evaluate (n^2): {(a, b), (a, c), (b, d), (e, c), (a, h)}.

3. Determine if the function $g(z) = 5z^4 + z^3 + 16$ is odd, even, or neither. Then evaluate g(- 1).

4. For the function $f(t) = \frac{(6t^2 + t^4)}{t^2}$, evaluate [2 + f(- 3)].

   (7-16-20-17)

As a cure for worrying, work is better than whiskey.

Thomas A. Edison

1. Graph the relation $x = y^2 - 9$. If the point (x, 4) is on the graph of this relation, find x.

2. If the point (x, - 2) is on the graph of the relation $x = - 2y^2 + 25$, find x.

3. Graph the function using graphing techniques, and then determine the lowest value in the range of the function: $f(x) = |x - 12| + 20$.

4. For the function $g(x) = 3 + \sqrt{(- x^2 + x + 56)}$, find the length of the interval on the x-axis that represents its domain.

   (7-17-20-15)

We want to live in the present, and the only history that is worth a tinker's damn is the history we make today.

Henry Ford

1. If the point (3, b) is on the graph of the relation $x^2 + y^2 = 25$, find the positive value of b, then evaluate (3 + b).

2. If point (6, c) is on the graph of the relation $x^2 - 4x + 6 = y$, find c.

3. Graph the function $f(t) = 2(t - 2)^2 + 12$ by using graphing techniques, then find its y-intercept.

4. If $f(x) = 3\sqrt{-x} + |x| - 3$, determine the domain of f(x) and then evaluate   f(- 9).

(7-18-20-15)

We must make the choices that enable us to fulfill the deepest capacities of our real selves.

Thomas Merton

1. If $f(x) = x^2 - 5x + c$ and f(2) = 1, find c.

2. Given that $g(t) = -t^2 + 3t + 59$, find g(- 5).

3. Graph the parabola $y = -3x^2 + 6x + 17$, and find its maximum value of y.

4. Graph the function $g(x) = -\sqrt{x - 7} + 7$. Then find (b + 1) if the point (b, 4) is on the graph of g(x).

(7-19-20-17)

Reason, ruling alone, is a force confining; and passion, unat-
tended, is a flame that burns to its own destruction.

Kahlil Gibran

1. For the function $h(z) = -\frac{3}{2}z + c$, find c if the point (2, 4) is on the graph of the function.

2. Find b, if the point (9, b) is on the graph of $f(x) = \frac{5}{3}x + 5$.

3. Graph the function $f(x) = \left(x - \frac{9}{2}\right)^2 - \frac{1}{4}$, and evaluate the product of its x-intercepts.

4. Find b if the point (- 3, b) is on the graph of $g(z) = -2|z| + 4z + 33$.

(7-20-20-15)

People can have many different kinds of pleasure. The real one is that
for which they will forsake the others.

Marcel Proust

1. The point (8, 13) is on the graph of the function $g(t) = \frac{5\,t-c}{t-6}$. Find t if f(t) = 21.

2. If the point (10, 9) is on the graph of the function $f(x) = \frac{b-5x}{6-x}$, find f(7).

3. For the function $f(x) = \sqrt{3 + |3 - x|} + x + 8$, find f(9).

4. Graph the parabola: $g(x) = -(x + 3)^2 + 24$, find its vertex, and then indicate its y-intercept.

(7-21-20-15)

This – the immediate, everyday, and present experience – is IT, the entire and ultimate point for the existence of a universe.

Alan Watts

1. Find the zero of the function $g(x) = -\frac{3}{5}x + \frac{21}{5}$.

2. Given that (11, 2) is a point on the graph of $h(t) = -\frac{2}{11}t + b$, find the zero of the function.

3. If $f(x) = 2x - 2$ and $g(x) = x^2 + 2$, evaluate the composition $(f \circ g)(3)$.

4. Given $g(t) = -2t - 1$ and $h(t) = t^3$, evaluate the composition $(g \circ h)(-2)$.

(7-22-20-15)

Nothing great was ever achieved without enthusiasm.

Ralph Waldo Emerson

1. Find the y-intercept of the linear function $f(x) = -\frac{3}{2}x + b$ if its zero is $4\frac{2}{3}$.

2. Given that the zero of the function $g(z) = 2z + b$ is -12.5, and that (- 1, c) is on the graph of $g(z)$, find c.

3. If $f(x) = -3x - 1$, evaluate the composition $(f \circ f)(2)$.

4. Given $g(t) = t^2$, find $[(g \circ g)(2) - 1]$.

(7-23-20-15)

Too many people overvalue what they are not and undervalue what they are.

Malcolm Forbes

1. Solve the compound inequality, graph the solution on a number line and find the number of integers that belong to the solution:
$2x - 8 < 0$ and $-2x - 7 \le 2 + x$.

2. Find the length of the interval on the number line that is not a solution of the compound inequality:
$x + 34 < -2 - x$ or $3x - 19 > x - 7$.

3. If $f(x) = 2x - 8$ and $g(x) = 3x^2 + 2$, find the composition $(f \circ g)(-2)$.

4. For $f(x) = 2x - 8$, find $(f(f(x)))$, then evaluate $[(f(f(10)) + 1]$.

(7-24-20-17)

Great is the road I climb, but ... the garland offered by an easier effort is not worth the gathering.

Propertius

1. Find the largest value of x that is a solution of: $-\frac{1}{2} < \frac{x+5}{2} \le 6$.

2. Find the smallest integer in the solution of: $5 \le \frac{x-5}{4} < 10$.

3. If $f(x) = \sqrt{(x^2 - 176)}$ and $g(x) = x^2 - 1$, find the composition of f and g, then evaluate your answer for $x = 5$.

4. Given $h(t) = \frac{3t}{t+2.4}$ and $k(t) = 2t - 1$, evaluate the composition of h and k for $t = -1$.

(7-25-20-15)

There is one thing certain, namely, that we can have nothing certain;
therefore it is not certain that we can have nothing certain.
Samuel Butler

1. Solve the absolute value equation and identify the positive solution:
$$|-3x + 6| = 15.$$

2. If m and n are the solutions of the absolute value equation $|2x - 21| = |x|$, evaluate $(m + n - 2)$.

3. If $f(x) = \frac{x-3}{x+1}$ and $g(x) = \frac{2}{x}$, find the composition of f and g, determine its domain, and find x such that the composition is equal to $-\frac{29}{11}$.

4. Let $f(t) = \sqrt{3t}$ and $g(t) = 2 + t$. Find the domain of the composition of f and g and then find t if $(f \circ g)(t) = \sqrt{51}$.

(7-26-20-15)

What a man thinks of himself, that is what determines,
or rather indicates his fate.
Henry David Thoreau

1. Find the largest solution of the absolute value equation:
$-|3x - 18| + 2 = -1.$

2. If b and c are the solutions of the equation in problem #1, evaluate: $2(b + c) + 3.$

3. The linear function F(C) converts Celsius degrees into Fahrenheit degrees. If F = 68 when C = 20, and F = 50 when C = 10, find F(C) and determine the value of C such that $(F \circ F)(C) = \frac{772}{5}$.

4. For the function F(C) from problem #3, find C if F(C) = 59.

(7-27-20-15)

That is the principal thing: not to remain with the dream, with the inten-tion, with the being in the mood, but always forcibly to convert it into all things.

Rainer Maria Rilke

1. Find the upper limit of the interval that is the solution of the ab-solute value inequality: $|2x - 3| \leq 11$.

2. Find the absolute value of the product of the left- and right-side limits of the solution interval found in problem #1.

3. Sketch a graph the given quadratic function by locating the ver-tex and the direction of opening. Then determine the y-inter-cept: $y = 2(x - 1)^2 + 18$.

4. Find b if the point (4, 34) is on the graph of $f(x) = (x - 1)^3 + b$. Then graph the function by transformations and find $[2 + f(3)]$.

(7-28-20-17)

He who postpones the hour of living is like the rustic who waits for the river to run out before he crosses.

Horace

1. Find the smallest integer number that cannot be a solution of the absolute value inequality: $|-2x + 15| > 1$.

2. If n is the largest integer that cannot be a solution of the ine-quality in problem #1, evaluate $(4n - 3)$.

3. Graph the parabola $g(x) = -x^2 + x + 22$, then evaluate: $|g(-6)|$.

4. Graph the function by transformations, find its x-intercept(s), and then locate the y-intercept: $h(x) = -(x - 1)^4 + 16$.

(7-29-20-15)

Imagination has always had powers of resurrection that no science can match.

Ingrid Bengis

1. If y varies directly with x and inversely with z, and y = 1 when x = 1 and z = 3, find y when x = 14 and z = 6.

2. Under the conditions from problem #1, find x when y = 45 and z = 2.

3. Graph the function $f(t) = -t^2 + 6t + 11$. Then find the maximum value of f(t).

4. Graph the parabola $g(z) = 2z^2 + 8z + 23$, then find the lowest value of g(z).

(7-30-20-15)

All business proceeds on beliefs, or judgment of probabilities, and not on certainties.

Charles W. Eliot

1. If y varies inversely with the square of x, and y = .5 when $x = 2\sqrt{14}$, find y when x = 2.

2. If y varies jointly with x and z, and y = 30 when x = 3 and z = 5, evaluate: $(y - \sqrt{y} + 1)$, if x = 6 and z = 3.

3. For the function $f(x) = x^2 + 8x - 84$, determine the length of the interval on the x-axis where $f(x) \le 0$.

4. For the parabola $y = -x^2 + 3(x + 18)$, determine the length of the interval on the x-axis where $y \ge 0$.

(7-31-20-15)

The meaning of things lies not in the things themselves, but in our atti-tude towards them.

Antoine de Saint-Exupery

1. Find x from the system of equations:
   $$3x - 2y = 22$$
   $$x + 3y = 11$$

2. From the system given in problem #1, find the value of y.

3. Graph the polynomial function and find the absolute value of the product of its nonzero x-intercepts: $f(x) = x^3 - x^2 - 20x$.

4. Sketch a graph for the polynomial function $g(t) = (t + 1)(t^2 - 2t - 15)$. Find the product of the x-intercepts and show that it is equal to the absolute value of the y-intercept.

(8-1-20-15)

Associate yourself with men of good quality if you esteem your own rep-utation, for 'tis better to be alone than in bad company.

George Washington

1. Solve the system of linear equations for x: $2x - 17 = -.5y$
   $$1.5x - y = 10$$

2. From the system given in problem #1, find y.

3. Sketch a graph of the function $f(x) = -3x(x + 5)((x - 4)(x + 1)$ by finding its x-intercepts and the starting and ending quadrants. Then evaluate the product of the nonzero x-intercepts.

4. Evaluate the expression: $[2 + \frac{-1}{60}f(5)]$, if f is the function from problem #3.

(8-2-20-17)

When there is no feeling of accomplishment, children fail to develop properly and old people rapidly decline.

Joseph Whitney

1. Two angles are complementary. If the difference between twice the smaller angle and the larger is 33, find the difference between the larger and the smaller.
2. Two angles are supplementary. The larger angle minus twice the smaller is 45. How many times is one angle larger than the other?
3. Sketch a graph of the rational function by determining the x-intercepts and the asymptotes, then find the y-intercept:
    $f(x) = \dfrac{((x-5)(x+4))}{x-1}$.
4. Find the x-intercept, the asymptotes, and sketch a graph of the rational function $g(x) = \dfrac{x-1}{x+3}$. Then evaluate $[g(-3\frac{2}{7}) + 2]$.

(8-3-20-17)

Unless a man has been taught what to do with success after getting it, the achievement of it must inevitably leave him a prey to boredom.

Bertrand Russell

1. Angles x and y are complementary, with x being the smaller angle. Their difference is 70. Evaluate $\dfrac{y}{x}$.
2. Angles m and n are supplementary. If 6m is 72 more than n, evaluate $\dfrac{n}{m}$.
3. Find the sum of the t-intercepts of the vertical asymptotes of the function:
    $G(t) = \dfrac{t^2}{(t^2 - 20t + 19)}$.
4. Find the intercepts, the asymptote(s), and sketch a graph of the function:
    $H(t) = \dfrac{(t^2 - 6t + 9)}{(t^2 + 9)}$. Then evaluate $(2 + 5x)$, where x is the x-intercept.

(8-4-20-17)

I like the dreams for the future better than the history of the past.
Thomas Jefferson

1. KK has $2.50 in dimes and quarters. If the number of quarters is 7 less than 3 times the number of dimes, how many quarters does KK have?

2. In problem #1, how many dimes does KK have?

3. Let k be the positive value of x for which the function $f(x) = \frac{(x^2-9)}{(x^2-2x-3)}$ is undefined. Evaluate $(6\frac{2}{3}k)$.

4. For the function from problem #3, let b be the y-intercept. Evaluate $(5b + 2)$.

(8-5-20-17)

Pray not for lighter burdens, but for stronger backs.
Theodore Roosevelt

1. The difference of two numbers is 2, and their sum is 14. Find the larger number.

2. One number is 14 less than another. Their sum is 26. Find the smaller number.

3. How many non-zero integers are there in the solution of the following inequality:
$x^2 + 4x - 96 \leq 0$?

4. Find the length of the interval on the number line that is not part of the solution of the inequality: $x^2 - 9x - 36 > 0$.

(8-6-20-15)

Work and acquire, and thou hast chained the wheel of Chance.
Ralph Waldo Emerson

1. John is two years younger than Mary. Five years from now the sum of their ages will be 42. Find one fourth of the sum of their ages now.

2. Cory and Day inherit $117,000. Two times the amount of Cory is $41,000 less than three times the amount of Day. Find *the number of thousands* Cory's amount is larger than that of Day.

3. Find the length of the solution interval of the inequality: $x(x-6) \le 91$.

4. Find the smallest positive value of x such that: $x^2 - 12x \ge 45$.
   (8-7-20-15)

One thing at a time, all things in succession. That which grows slowly endures.
J.G. Holland

1. Graph the system of inequalities and find the x-coordinate of the corner point:
$$3x - y \le 16$$
$$x + 2y \le 24$$

2. In problem #1, what is the y-coordinate of the corner point of the solution?

3. Solve the inequality and find the product of the smallest two positive integers that cannot be in its solution: $\dfrac{x+1}{(x^2 - 9x + 20)} \ge 0$.

4. Find $|ab|$ if [a, b] is the solution interval of the inequality:
$$\frac{(x^2 - 2x - 15)}{(x^2 - 6x + 9)} \le 0.$$
   (8-8-20-15)

Every minute of life carries with it its miraculous value,
and its face of eternal youth.

Albert Camus

1. Graph the system of inequalities and find the y-coordinate of the corner point(s) not on the x-axis: $8x + 9y \geq 72$, $x \leq 9$ and $y \leq 8$.

2. What is the x-coordinate of the corner point of the solution from problem #1, that is not located on the axes?

3. Find the absolute value of the largest integer that is not a solution of the inequality: $\frac{-12}{2x+28} < 1$.

4. Find the largest value of x for which: $\frac{x-8}{2x+5} \leq \frac{1}{5}$.

(8-9-20-15)

Were it not for my little jokes, I could not bear the burdens of this office.

Abraham Lincoln

1. For the functions $f(x) = 3x - 2$ and $g(x) = 2x^2 + x + 4$, find $(f + g)(1)$.

2. For the function $h(t) = 3t^2 + 3$, evaluate $\frac{1}{3}h(-3)$.

3. Find the product of all non-zero zeros of the polynomial function:
   $f(x) = 2x^3 - 18x^2 + 40x$.

4. Knowing that $x = i\sqrt{2}$ is a complex zero of the polynomial function with real coefficients $f(x) = x^4 - 8x^3 + 17x^2 - 16x + 30$, find the product of all the real zeros of the function.

(8-10-20-15)

He is happy whose circumstances suit his temper; but he is more excel-
lent who can suit his temper to any circumstances.

David Hume

1. For the functions $g(z) = -2z^2 + 3z$ and $h(z) = z - 4$, find $(g - h)(z)$, and then find the absolute value of $(g - h)(-2)$.

2. For the functions in problem #1, evaluate $[-9 + (h - g)(4)]$.

3. Given the real polynomial function $f(x) = x^4 - 12x^3 + 47x^2 - 60x$ and that $x = 3$ is one of its zeros, find the product of the other non-zero zeros.

4. If t, $t + 1$, $t + 2$ are the non-zero zeros of the function from problem #3, evaluate $t(t + 2)$.

(8-11-20-15)

There is only one corner of the universe you can be certain of improving,
and that's your own self.

Aldous Huxley

1. If $f(x) = x^2 + 5$ and $g(x) = 2x^2 - 3x + 1$, evaluate $[2 + (g - f)(5)]$.

2. For the functions $f(t) = 2t^3 - 4$ and $g(t) = -t^2 + 2t$, evaluate $(f + g)(2)$.

3. If three of the zeros of a real fourth degree polynomial function with the leading coefficient equal to 1 are 2, -2, and $1 - 2i$, find the absolute value of its y-intercept.

4. If $f(t)$ is the function from problem #3, evaluate $[2 + \frac{3}{8}f(3)]$.

(8-12-20-17)

He who would make serious use of his life must always act as though he had a long time to live and schedule his time as though he was about to die.

Emile Littre

1. Multiply and simplify: $(x + y + 1)(x + y - 1)$. Then evaluate your answer for $x = 2$ and $y = 1$.

2. Find the positive value of z such that the product $(z - 3)(z + 3)$ is equal to 160.

3. If a 4th degree real polynomial function with the leading coefficient 1 has only complex zeros and two of them are $(2i)$ and $(- i\sqrt{5})$, sketch a graph of the function, and indicate its y-intercept.

4. For the function from problem #3, evaluate $[\frac{1}{2}f(- 1) + 2]$.

(8-13-20-17)

He who postpones the hour of living is like the rustic who waits for the river to run out before he crosses.

Horace

1. Multiply and simplify: $[3 - (a - b)][3 + (a - b)]$. Then, evaluate your answer for $a = - 1$ and $b = - 2$.

2. Multiply and then simplify the express $E = (x^2 + 2x + 4)(x - 2)$. Then, evaluate $(E + 30)$ for $x = - 2$.

3. If the four zeros of a 4th degree polynomial function with leading coefficient 1 are ± 2 and $\pm\sqrt{5}$, find the function, sketch its graph, and indicate which coefficient of the polynomial shows the product of all the zeros.

4. For the function from problem #3, evaluate: $[\frac{f(-3)}{5} + 13]$.

(8-14-20-17)

When you rise in the morning, form a resolution to make the day a happy one for a fellow creature.

Sydney Smith

1. For the function $f(x) = -2x^2 + x + 18$, evaluate $f(-2)$.

2. For the function $g(x) = x^2 - 2x$, find $g(x + h) - g(x)$, and then evaluate your answer for $x = 8$ and $h = 1$.

3. Find the largest potential rational zero (t) of the polynomial function $f(x) = 2x^3 - 17x^2 + 31x - 16$, then evaluate $(1.25t)$.

4. For the function from problem #3, find the double of the sum of the irrational zeros.

(8-15-20-15)

Man must cease attributing his problems to his environment and learn again to exercise ... his personal responsibility in the realm of faith and morals.

Albert Schweitzer

1. In the function $h(t) = -t^3 + k$, find the value of k if $h(2) = -8$, and then evaluate $h(-2)$.

2. For the function $f(x) = 2x^2 - x$, find $f(x + h) - f(x)$, and then evaluate your answer for $x = 3\frac{3}{4}$ and $h = 1$.

3. If $-\frac{1}{2} - \frac{\sqrt{3}}{2}i$ is a complex zero of the real polynomial function $g(x) = x^4 + x^3 - 63x^2 - 64x - 64$, find the largest rational zero of the polynomial (z) and evaluate $(2.5z)$.

4. Graph the function from problem #3, and evaluate the sum (s) of the absolute values of the real zeros. Then find $(s + 1)$.

(8-16-20-17)

A man should remove not only unnecessary acts, but also unnecessary thoughts, for then superfluous activity will not follow.

Marcus Aurelius

1. Divide and then evaluate your answer for x = 1: $\dfrac{\left(4x^4+8x^3+4x^2\right)}{2x^2}$.

2. If Q(x) is the quotient of $\dfrac{\left(3x^2-4x-15\right)}{x-3}$, evaluate Q(4).

3. For the real polynomial function h(t) = t³ − 18t² + 15t + 34, find all its rational zeros, graph the function, and evaluate (z + 3) if z is its largest rational zero.

4. If x = - i√2 is a complex zero of the real polynomial function g(x) = x³ − 15x² + 2x − 30, find its rational zero.

(8-17-20-15)

Time is the most valuable thing a man can spend.

Theophrastus

1. Evaluate the remainder of the following division for
   x = - 1: $\dfrac{\left(4x^3-6x^2+3\right)}{2x^2+1}$.

2. If D(x) = x³ − 8 and d(x) = x² + 2x + 4, evaluate $\dfrac{D(x)}{d(x)}$ for x = 20.

3. Use the Intermediate Value Theorem to show that f(x) = x³ − x² +2x − 2 has a real zero in the interval [- 2, 2], find that zero (t) then evaluate (t² + 15t + 4).

4. If p and q are the complex zeros of the function from problem #3, evaluate (p⁴ + q⁴ + 9).

(8-18-20-17)

Children today are tyrants. They contradict their parents, gobble their food, and tyrannize their teachers.

Socrates (470-399 B.C.)

1. Use synthetic division to find the remainder: $\dfrac{(3x^3 - 2x^2 + x + 42)}{x+2}$.

2. If Q(x) is the quotient of the division from problem #1, evaluate $[Q(3) - 1]$.

3. Two of the real zeros of a 4th degree polynomial function are ± 2. The sum and the product of the other two zeros are (- 4) and (- 5) respectively. Sketch a graph of the function and indicate its y-intercept.

4. Use the Intermediate Value Theorem to show that the real polynomial function $f(x) = x^4 - 8x^3 + 18x^2 - 24x + 45$ has a zero in the interval [2, 4] and find the real zeros knowing that $x = -i\sqrt{3}$ is one of its complex zeros. Then, determine the product of the real zeros.

(8-19-20-15)

Let others praise ancient times; I am glad I was born in these.

Ovid (81 B.C.)

1. Use synthetic division to find the remainder of the division: $\dfrac{(4x^4 - 2x^2 - 48)}{x = 2}$.

2. If Q(x) is the quotient of the division from problem #1, find; $|Q(1) - 2|$.

3. A 4th degree polynomial function f(x) with real coefficients has zeros 4, i, and 5. Find the function, sketch its graph, and indicate its y-intercept.

4. For the function from problem #3, evaluate (.5f(2) + 2).

(8-20-20-17)

He that will not sail till all dangers are over must never put to sea.
Thomas Fuller

1. Factor the polynomial $2x^3 + 12x^2 + 10x$ completely, then evaluate each factor for $x = 3$ and indicate the largest factor.

2. Factor the polynomial function $P(x) = 3x^3 - 2x^2 - 9x + 6$ completely. Then use the factored form to evaluate $(.5P(3))$.

3. A polynomial function of 5th degree with real coefficients has zeros $1 + i$, -5, and ± 2. Find the function and the product of all its zeros (P) and evaluate $(.5P)$.

4. A real polynomial function of 4th degree with the first coefficient 1, has zeros i, 3, and 5. Find the function, sketch its graph, and indicate its y-intercept.

(8-21-20-15)

He who has never failed somewhere, that man cannot be great.
Herman Melville

1. Factor the trinomial $3z^2 + 10z - 8$ and indicate the value of the smallest factor different than 1 when $z = 4$.

2. Factor $4y^2 - 2y - 8$, then evaluate your answer for $y = 3$.

3. Solve the equation in the complex number set and indicate the product of all the solutions: $x^4 + 9x^3 + 21x^2 = -9x - 20$.

4. Solve the equation and find the product of its solutions: $2x(x^2 - 13) = 6(5 - x^2)$.

(8-22-20-15)

With begging and scrambling we find very little, but with being true to ourselves we find a great deal more.

Rabindranath Tagore

1. Factor $x^3 - 27$ over real numbers, and evaluate the smaller factor for x = 11.

2. Factor $t^3 + 64$ over real numbers and evaluate the smaller factor for t = 19.

3. The diagonal of a rectangle is 50 in. Its perimeter is 140 in. Evaluate half of its longer side.

4. For the rectangle in problem #3, evaluate half of the shorter side of the rectangle.

(8-23-20-15)

A man should not strive to eliminate his complexes but to get into accord with them, for they are legitimately what directs his conduct in the world.

Sigmund Freud

1. Factor completely: $2t^3 - 3t^2 - 8t + 12$. Then find the absolute value of the sum of the smaller two factors when t = - 1.

2. Factor completely as a difference of squares: $y^6 - 64$. Then, if k is the largest factor when y = 3, evaluate (k + 5).

3. For the one-to-one function f(x) = $\frac{2x-3}{x+1}$, find its domain, and evaluate $[-\frac{10}{3}f^{-1}(3)]$.

4. If g(x) = $\frac{2}{x+1}$, evaluate $[- 25g^{-1}(5) + 2]$.

(8-24-20-17)

Obstacles cannot crush me, every obstacle yields to stern resolve.

Leonardo da Vinci

1. Factor $y^4 - y^2 - 12$ completely and find the value of the smallest factor if $y = 10$.

2. Factor $\frac{1}{4} - \frac{x^2}{9}$ as a difference of squares and find the value of the larger factor when $x = 73\frac{1}{2}$.

3. If $f(t) = \sqrt{t - 6}$ find its domain, then evaluate $[f^{-1}(4) - 2]$.

4. For $f(x) = x^{\frac{1}{3}} - 5$, evaluate $[9 + f^{-1}(-3)]$.

(8-25-20-17)

My formula for living is quite simple. I get up in the morning and I go to bed at night. In between I occupy myself as best I can.

Cary Grant

1. If $G(x) = x^2 - 6x - 5$, find the positive value of x for which $G(x) = 11$.

2. If $f(z) = z^2 - 11z - 9$, find the absolute value of the product of the solutions of the equation $f(z) = 17$.

3. Find the exact value of the expression; $3 + \log_2 64 + \ln e^{11}$.

4. Evaluate: $2 + \log 10^3 - \log_3 27 + 2^{\log_2 15}$.

(8-26-20-17)

The man is happiest who lives from day to day and asks no more, garnering the simple goodness of a life.

Euripides

1. Find the largest value of x for which f(x) = 4, if f(x) = $x^2 - 6x - 12$.

2. For g(x) = $-x^2 + 4x + k$, find k such that g(-5) = -18.

3. Simplify the expression and evaluate it for
   x = 8 and y = 81: $\log_2 x^4 + \log_3 y^2$.

4. Simplify the expression and evaluate it for
   t = e^3 and p = 1000: $\ln t^3 + \log p^2$.

(8-27-20-15)

If I lose, I'll walk away and never feel bad ... Because I did all I could, there was nothing more to do.

Joe Frasier

1. Find the domain of F(x) = $\dfrac{(x^2 - 4)}{(x^2 - x - 6)}$, then find $\dfrac{F(x)}{G(X)}$ if
   G(x) = $\dfrac{x+2}{x^2 + x - 12}$, and simplify. What is 2 times the absolute value
   of the product of the solutions of the equation: $\dfrac{F(x)}{G(x)} = 0$?

2. For the $\dfrac{F(x)}{G(X)}$ from problem #1, evaluate $[-\dfrac{28}{9} \cdot \dfrac{F(x)}{G(X)}]$ for x = -1.

3. Write the expression $\log_2\left(a\sqrt{b}\right)^2$ as a sum of single logarithms,
   then evaluate it for a = 64 and b = 256:

4. Write the expression $\log\left(x^2\sqrt{y}\right)^3$ as a sum of single logarithms,
   then evaluate it for x = y = 100:

(8-28-20-15)

He has not learned the first lesson of life who does not every day surmount a fear.

Ralph Waldo Emerson

1. For the functions $k(t) = \frac{t-3}{(t^2)-4}$ and $h(t) = \frac{(t^2-9)}{t+2}$, find $Q(t) = \left(\frac{h}{k}\right)(t)$ and simplify. Then evaluate; $Q(-2) + 12$.

2. For $Q(t)$ from problem #1, evaluate $[Q(-7) - 7]$.

3. Simplify the expression to a sum of single logarithms, then evaluate it for $x = 3$: $\log_2((x+5)^2(2x-2)^3(x+1)^4)$.

4. Simplify the expression to a sum of single logarithms and evaluate your answer when $x = e$: $\ln[\frac{((2x-e)^5(3x-2e)^8 x^3)}{e}]$.

(8-29-20-15)

Look well into thyself; there is a source which will always spring up if thou will always search there.

Marcus Aurelius

1. For the functions $f(x) = \frac{-x+3}{(x^2-3x-40)}$ and $g(x) = \frac{(x^2-9)}{(x^2-7x-8)}$ find $h(x) = \frac{g}{f}(x)$. Then evaluate $[h(-2) + 5]$.

2. For $h(x)$ from problem #1, evaluate $[\frac{-9}{h(-2)} + 33]$.

3. Write the expression as a single logarithm and evaluate it for $x = 3$: $2\log_3(x^2 + 18) + 3\log_3(2x + 3) + 2\log_3(x^3 + 54)$.

4. Write the expression as a single logarithm and evaluate it for $x = 10$: $3\log(3x + 70) + 4\log(4x + 60) - \frac{1}{3}\log(x - 9) + \frac{1}{2}\log(x + 90) + 2$.

(8-30-20-17)

They are happy men whose natures sort with their vocations.

Francis Bacon

1. For the functions $f(t) = \dfrac{t-2}{(t^2 + 2t + 4)}$ and $g(t) = \dfrac{(t^3 - 8)}{t+1}$, find $h(t) = (f \cdot g)(t)$. Then evaluate $[-3h(2) + 8]$.

2. For $h(t)$ from problem #1, evaluate $[-h(-4) + 19]$.

3. Find y as a function of x if c is a positive non-negative constant: $\ln y = 2x + \ln c$. Then evaluate y for $x = \ln 2$ and $c = 5$.

4. Write y as a function of x, then evaluate $(2y + 1)$ for $x = \sqrt{\ln 4}$: $\ln(y - 2) + \ln 2 = x^2 + \ln 3$.

(8-31-20-17)

Experience is not what happens to you; it is what you do with what happens to you.

Aldous Huxley

1. Perform the operations and simplify: $\dfrac{x+5}{(x^2 - 4)} + \dfrac{x}{x-2}$. Then evaluate the numerator of your answer for $x = 1$.

2. For problem #1, evaluate $(S + 1)$, where S is the sum of the numbers for which the answer is undefined.

3. Find the exact value of the expression: $E = \log_3 \dfrac{1}{243}$. Then evaluate $(-4E)$.

4. Find the exact value of: $(2 + 5\ln e^3)$.

(9-1-20-17)

Simplicity, simplicity, simplicity! I say, let your affairs be as two or three,
and not a hundred or a thousand ... Simplify, simplify.

Henry David Thoreau

1. Subtract the fractions and simplify. Then evaluate the numerator for x = - 2: $\frac{3x-2}{(x^2+3)} - \frac{1}{x}$.

2. Perform the operations and simplify: $\frac{3}{x-1} - \frac{2}{x+2} - \frac{1}{x}$. Then evaluate
   the numerator for x = 0.

3. Evaluate: $5(2^{\log_2 4})$.

4. Evaluate: $e^{\ln 6} + 3\log_3 27 + 2$.

(9-2-20-17)

What you can do, or dream you can do, begin it; boldness has genius,
power and magic in it.

Johann von Goethe

1. Simplify the expression: $\frac{y+3}{(y^2-y-12)} - \frac{y}{(y^2-16)}$. Then evaluate
   (N + 5), where N is the value of the numerator of your answer
   for   y = -1.

2. For problem #1, find the absolute value of the negative number
   that makes the first fraction undefined.

3. Graph the function by hand: $f(x) = 2^{x-3} + 4$. Then evaluate f(7).

4. Graph g(x) and indicate the y-intercept: $g(x) = \frac{1}{3} \cdot 3^{-x+3} + 6$.

(9-3-20-15)

We can do anything we want to do if we stick to it long enough.
Helen Keller

1. Simplify, without negative exponents in the final answer: $\frac{(x^{-2}-y^{-2})}{(x^{-1}+y^{-1})}$. Then evaluate the numerator of your answer for x = - 3 and y = 6.

2. If N is the expression in simplest form from problem #1, evaluate $\frac{1+N}{N}$ for x = 2 and y = 6.

3. Graph the function, then find f(e²): f(x) = 12 + 4lnx.

4. Graph the function g(t) = $4e^{2t}$ – 1. Then evaluate [g(ln2) + 2].

(9-4-20-17)

Failure is only an opportunity to begin again more intelligently.
Henry Ford

1. Simplify the expression (no negative exponents in the final answer). Then evaluate it for x = 2 and y = 3: $\frac{(3x^{-1}+3y^{-1})}{(2x^{-1}-2y^{-1})}$ + 1.5.

2. Evaluate (P + .5) for x = 1 and y = 2, where P is the simplest form of the expression in problem #1.

3. If S is the solution of the equation 4^{1-2x} = 2, evaluate (80S).

4. Solve the equation: $\frac{1}{3^{x-2}}$ = 3^{x-4}. Then find the product of 5 and the solution of the equation.

(9-5-20-15)

The man with a toothache thinks everyone happy whose teeth are sound. The poverty-stricken man makes the same mistake about the rich man.

George Bernard Shaw

1. For the functions $f(x) = \frac{3x}{x+1}$ and $g(x) = \frac{x+2}{2x}$, find the positive value of t for which $f(t) = g(t)$. Then evaluate $(9t)$.

2. Evaluate $(6t)$, if t is the largest value of t for which $f(t) = g(t)$ in problem #1.

3. Solve for t and then evaluate $(4t)$: $\qquad$ $8^{5+t} = 4^{3t}$.

4. Solve the equation: $5^{-2x-3} - 25 = 100$. Then find the product of (-5) and the solution of the equation.

(9-6-20-15)

He who does not get fun and enjoyment out of every day ... needs to re-organize his life.

George Matthew Adams

1. Let k be the value of t for which $g(t) = h(t)$ when $g(t) = \frac{t-1}{t-2}$ and $h(t) = \frac{t}{t+2}$. Then evaluate $(13.5k)$.

2. Solve the equation: $\frac{2}{x+3} + \frac{3}{x-3} = \frac{38}{x^2-9}$.

3. If t is the solution of the equation $2^{2x-1.5} = \sqrt{32}$, evaluate $(10t)$.

4. Solve the equation: $3^{x^2+x} - 81^3 = 0$. Then find the product of 5 and the positive solution of the equation.

(9-7-20-15)

The happiest excitement in life is to be convinced that one is fighting for all one is worth on behalf of some clearly seen and deeply felt good.

Ruth Benedict

1. Solve the rational equation: $\dfrac{2x-1}{x+7} - \dfrac{3}{5} = \dfrac{37}{5x+35}$.

2. Karla and Leslie can paint a room together in $3\frac{3}{7}$ hours. Leslie alone can paint that room in 6 hours. How long will it take Karla to paint the same room alone?

3. Solve the equation and find the product of 2.5 and the positive solution: $\log_2(x^2 - 6x) - 4 = 0$.

4. Solve the equation: $\log_{\sqrt{2}}(x - 7) - 6 = 0$.

(9-8-20-15)

There is no more miserable human being than one in whom nothing is habitual but indecision.

William James

1. The formula for the area of a trapezoid is $A = \dfrac{h}{2}(b + B)$. Solve the formula for B, then find B when h = 9, b = 5, and A = 63.

2. In the equation from problem #1, solve for h, then evaluate h when A = 81, b = 7, and B = 11.

3. If s is the solution of the equation $3^{x-12} \cdot 9^x = 1$, find (5s).

4. Find the solution t of the equation, then evaluate (3.75t + 2): $9^{x+2} - 27^x = 0$.

(9-9-20-17)

One can never consent to creep when one feels an impulse to soar.

Helen Keller

1. Solve the equation for k, and find k when C = 10 and
   F = 50: $C = \frac{5}{k}(F - 32)$.

2. The equation $F = \frac{9}{5}C + 32$ is used to convert Celsius into Fahrenheit degrees. Solve the equation for C and find C when F = 50.

3. Find the solution of the equation $2^x \cdot 625 = 10^x$, and then evaluate the product of 5 and the solution.

4. If t is the solution of the equation $32 - \frac{4^x}{2} = 0$, find $(5t + 2)$.

(9-10-20-17)

Courage and perseverance have a magical talisman, before which difficulties disappear, and obstacles vanish into air.

John Quincy Adams

1. One number is 6 less than another. The ratio of the numbers is 3. Find the larger number.

2. The ratio of 3 less than a number and 3 more than the same number, is $\frac{4}{7}$. Find the number.

3. Find (10t) if t is the solution of the equation:
   $\log_2(x - 1) = 1 - \log_2 x$.

4. If k is the solution of the equation
   $\ln(x + 5) - \ln x = \ln 2$, evaluate $(2 + 3k)$.

(9-11-20-17)

There are risks and costs to a program of action. But they are far less than the long-range risks and costs of comfortable inaction.
John F. Kennedy

1. Mark rides his bike 3 mph faster than George. If Mark would ride for 6 hours, he would have traveled double the distance George would cover in 4 hours. Find George's speed.

2. In problem #1, find Mark's speed.

3. If M is the multiple of π when you convert 720° to radians, find 5M.

4. Convert $\frac{\pi}{12}$ radians to degrees.

(9-12-20-15)

Begin somewhere; you cannot build a reputation on what you intend to do.
Liz Smith

1. If you increase each the numerator and the denominator of a fraction in simplest form by 7, the new fraction is equivalent to $\frac{4}{5}$. If instead you decrease them by 7, the new fraction is equivalent to $\frac{1}{3}$. Find the numerator of the original fraction.

2. Find the denominator of the original fraction in problem #1.

3. If $E = 2\tan\frac{\pi}{4} + 3\cot\frac{\pi}{4}$, find 4E.

4. For $F = \sin\frac{\pi}{3} + \cos\frac{\pi}{6}$, evaluate $(2 + 5F\sqrt{3})$.

(9-13-20-17)

Effort only fully releases its reward after a person refuses to quit.

Napoleon Hill

1. In the isosceles right triangle ABC with A = 90°, drop the following perpendiculars to the respective sides: AD to BC, DE to AB,
   DF to AC, FG to BC, and EH to BC. If BC = 36, find DH.

2. In problem #1, evaluate $\frac{7GH}{DH}$.

3. Find the precise value of the expression: 4(4cos 60° - 3tan 135°).

4. If E = 1 + $\sqrt{3}$ sec $\frac{\pi}{6}$ + $\sqrt{3}$ csc 120°, find the exact value of (3E + 2).

(9-14-20-17)

*We hear of a silent generation, more concerned with security than integ-
rity, with conforming than performing, with imitating than creating.*

Thomas J. Watson

1. For the function f(x) = $\frac{-x+B}{x-2}$ find B if f(- 5) = - 2.

2. If g(t) = $\frac{(-3t^2 - 2t + C)}{t - 5}$, find C when g(- 2) = - 1.

3. Find the precise value of E = 20 sin² $\frac{\pi}{9}$ + $\frac{20}{sec^2\ 20°}$.

4. Find the exact value of S = 2 + 6$\sqrt{3}$ sin $\frac{2\pi}{3}$ - 6$\sqrt{2}$ cos $\frac{3\pi}{4}$.

(9-15-20-17)

A life of reaction is a life of slavery, intellectually and spiritually. One must fight for a life of action, not reaction.

Rita Mae Brown

1. Find the value of p in the function $f(x) = \frac{(x^2 + 6x + P)}{4x + 12}$ if $f(5) = 2$.

2. In problem #1, with the value of p you found, simplify $f(x)$ and evaluate $f(61)$.

3. If $E = \frac{1}{cos^2 \, 40°} - \frac{1}{cot^2 \, 40°}$, find the exact value of 20E.

4. Simplify $P = \frac{-\tan(-10°)}{\tan 190°}$, then find the precise value of $(15P + 2)$.

(9-16-20-17)

The man with insight enough to admit his limitations comes nearest to perfection.

Johann von Goethe

1. Find and simplify the inverse (the reciprocal) of $27^{-\frac{2}{3}}$.

2. Evaluate: $81^{\frac{1}{2}} + 2^{\frac{6}{2}}$.

3. If $E = \sin 270° + \cos (-180°) + 3 \sin (-\frac{3\pi}{2})$, find the exact value of 20E.

4. Evaluate precisely: $2\sqrt{3} \sin \frac{2\pi}{3} - 4 \cos \frac{5\pi}{2} - 12 \sin \frac{7\pi}{2} + 2$.

(9-17-20-17)

148

Education is hanging around until you've caught on.

Robert Frost

1. Evaluate: $\dfrac{1}{8^{-\frac{5}{3}}}$ - 23.

2. Find the value of the expression for x = 27 and y = 16: $x^{\frac{2}{3}} \cdot y^{\frac{1}{4}}$.

3. If $\cos \theta = -\dfrac{3}{5}$ and $\sin \theta < 0$, evaluate $(15 \tan \theta)$.

4. Let $\csc \theta = \dfrac{5}{3}$ and $\cos < 0$. Find the exact value of $(2 - 12 \sec \theta)$.

(9-18-20-17)

The essence of philosophy is that a man should so live that his happiness shall depend as little as possible on external things.

Epictetus

1. Evaluate the expression $\sqrt{(x^4 y^3)}$ for x = 3 and y = 1.

2. If x = 2 and y = 9, find the value of $E = 2\sqrt{x^6} + \dfrac{1}{y^{-\frac{1}{2}}}$.

3. For $\tan \theta = -2$ and $\dfrac{3\pi}{2} < \theta < 2\pi$, find $20\sqrt{5} \cos \theta$.

4. If $\cot \theta = -2$ and $\dfrac{\pi}{2} < \theta < \pi$, evaluate: $2 + 15\sqrt{5} \sin \theta$.

(9-19-20-17)

Friendship with oneself is all-important, because without it one cannot be friends with anyone else.

Eleanor Roosevelt

1. In the 30° - 60° - 90° right triangle ABC, angle A = 90°. Drop the following perpendiculars to the respective sides: AD to BC, DE to AB, DF to AC, FG to BC, and EH to BC, then find GD if FE = $12\sqrt{3}$.

2. In problem #1, evaluate (2DH + 2).

3. Graph the function over one period: y = - 10 sin $(x - \frac{\pi}{2})$. Find the length of the interval that represents the range of y.

4. Graph the function over one period: y = - $\frac{1}{3}$ tan 2x + 15. Find the y-intercept.

(9-20-20-15)

Yesterday is a cancelled check; tomorrow is a promissory note; today is the only cash you have – so spend it wisely.

Kay Lyons

1. Simplify the expression: E = $\frac{\sqrt[4]{81}}{\sqrt[3]{64}}$. Then evaluate (12E).

2. Simplify: $\frac{27^{\frac{1}{3}}}{49^{\frac{-1}{2}}}$.

3. Graph the function and determine its amplitude: y = - 20 sin (x + 20°).

4. If T is the period of y = 2 cos (3πx), evaluate $\left(\frac{45T}{2} + 2\right)$.

(9-21-20-17)

We learn courageous action by going forward whenever fear urges us back.

David Seabury

1. If $E = \dfrac{\sqrt{6}}{\sqrt[4]{36}}$, find $(E + 8)$.

2. Find the exact value of the expression for $y = 64$:
$$\dfrac{\sqrt[3]{y^2}}{\sqrt{y}} \cdot 121^{\frac{1}{2}} .$$

3. Find the phase shift in degrees of the function:
$y = 3 \sin (2x - 40°) + 1.$

4. Find the phase shift in degrees of the function:
$y = -2 \cos (3x - \dfrac{\pi}{4}) + 4.$

(9-22-20-15)

No yesterdays are ever wasted for those who give themselves to today.

Brendan Francis

1. Simplify: $\dfrac{(\sqrt{27}\sqrt[3]{9})}{\sqrt[6]{3}} .$

2. Evaluate the expression for $x = 3$ and $y = 25$: $\sqrt[3]{9x} \sqrt{12x} + \sqrt[3]{5y} .$

3. For $y = -5 \cos (\pi x - 6) + 4$, find $(-4A)$, if A is its amplitude.

4. For the function from problem #3, find $(\dfrac{5\pi}{2} \cdot \omega + 2)$, where ω is its phase shift.

(9-23-20-17)

Hardship, poverty, and want are the best incentives, and the best foundation, for the success of man.

Bradford Merrill

1. Evaluate $\frac{(6+\sqrt{8x})}{2}$, for x = 18.

2. Evaluate the expression for t = 16: $\frac{(t+2\sqrt{16t})}{2}$.

3. If E = - tan (cos⁻¹ (- .5)) evaluate (E√3 + 17) for angles less than 180°.

4. Find $\frac{S}{2}$ if S = sin⁻¹ (cos $\frac{\pi}{3}$).

(9-24-20-15)

Nature gave men two ends – one to sit on, and one to think with. Ever since then man's success or failure has been dependent on the one he used most.

George R. Kirkpatrick

1. If E = $\frac{\sqrt[3]{-128x^8}}{-\sqrt[3]{2x^{-1}}}$, evaluate $\frac{E}{12}$ for x = 3.

2. Simplify the expression for x = 5 and y = - 5: $\frac{\sqrt{x^{-5}}}{\sqrt{-y^{-7}}} \cdot \sqrt{-xy}$

3. If E = cos²x (1 + tan²x), evaluate (E + 19).

4. Simplify S = sin x + $\frac{(cos^2x)}{1+\sin x}$. Then evaluate (S + 16).

(9-25-20-17)

A problem is a chance for you to do your best.

Duke Ellington

1. Evaluate the radical function $f(x) = \sqrt{-3x + 18}$ for x = - 21.

2. If $g(x) = 24 - \sqrt[3]{-3x - 2}$, find g(2).

3. Simplify: $E = \dfrac{1}{\csc x}\left(\dfrac{1-\cos x}{\sin x} + \dfrac{\sin x}{1-\cos x}\right)$. Then evaluate (10E).

4. Simplify: $T = -\cos^2 x + 3\sin^2 x + 4\cos^2 x$. Then evaluate (5T + 2).

(9-26-20-17)

From the discontent of man, the world's best progress springs.

Ella Wheeler Wilcox

1. In the 30° - 60° - 90° right triangle ABC, angle A = 90°. Drop the following perpendiculars to the respective sides: AD to BC, DE to AB, DF to AC, FG to BC, and EH to BC. If the area of the circle of diameter GH is 81π, find DH.

2. In problem #1, find HB.

3. If $E = \dfrac{1}{(\cos^2 x)}\left(1 - \dfrac{\left((2(\sin^2 x)- 1)^2\right)}{(\sin^4 x - \cos^4 x)}\right)$, evaluate (E + 18).

4. For $P = \dfrac{\cos(x-y)}{\cos x \cos y}$ - tan x tan y, evaluate (18 – P).

(9-27-20-17)

Employment ... is so essential to human happiness that indolence is justly considered the mother of misery.

Burton

1. Find h(25) if h(t) = - $\sqrt{t - 24}$ + $\sqrt{5t - 25}$.

2. If f(x) = 3 $\sqrt{x + 4}$ + $\sqrt[5]{x}$ + $\sqrt{2x}$, find f(32).

3. If A = cos x + sin x tan $\frac{x}{2}$, evaluate (20A).

4. Simplify T = $\frac{\sin(4x)}{\sin(2x)(1 - 2\sin^2 x)}$, and then evaluate (7.5T + 2).

(9-28-20-17)

You have to deal with the fact that your life is your life.

Alex Hailey

1. Find the largest integer in the domain of the function:
    g(x) = $\sqrt{18 - 2x}$.

2. Evaluate (g(x) + 25) for g(x) from problem #1, when x = 1.

3. Find the exact value of E = sin 75°, and then evaluate
    [($\sqrt{6}$ – $\sqrt{2}$)E + 19].

4. If P = - cos (-$\frac{5\pi}{6}$), find the precise value of (10P$\sqrt{3}$ + 2).

(9-29-20-17)

There are as many ways to live and grow as there are people. Our own ways are the only ways that should matter to us.

Evelyn Mandel

1. Find the domain of the function $f(x) = \sqrt{3(x + 27)}$. Then graph $f(x)$ and determine its y-intercept.

2. Graph the function $g(x) = -\sqrt{x + 6}$. Then find x such that $g(x) = -6$.

3. If $T = \tan \frac{\pi}{8}$, find the exact value of $[21 - (\sqrt{2} + 1)T]$.

4. Find the precise value of $A = \tan 15°$, and then evaluate $[(2 + \sqrt{3})A + 16]$.

(9-30-20-17)

Time can only be understood backwards, but it must be lived forward.

Soren Kierkegaard

1. Solve the equation: $\sqrt{x - 1} + \sqrt{x - 6} = 5$.

2. In the right triangle ABC, angle A = 90°. Drop the following perpendiculars to the respective sides: AD to BC, DE to AB, DF to AC, FG to BC, and EH to BC, and then evaluate $\frac{GD}{DH}$.

3. Angles x and y are in the first quadrant. If $\cos x = \frac{4}{5}$ and $\cos y = \frac{5}{13}$, find $T = \sin (x + y)$. Then evaluate $(\frac{1300T}{63})$.

4. With the given in problem #3, find $P = \tan (y - x)$ and then evaluate $(2 + \frac{280P}{11})$.

(10-1-20-17)

To the timid and hesitating everything is impossible because it seems so.
Sir Walter Scott

1. Solve the equation: $3\sqrt[3]{x-2}+2=8$.

2. Solve: $-2x^{\frac{1}{3}}+\sqrt[3]{16}=0$

3. Find the precise value of E = $\sin\left(\cos^{-1}\frac{5}{13}-\cos^{-1}\frac{4}{5}\right)$. Then evaluate $\left(E+\frac{1267}{65}\right)$.

4. Find the exact value of P = $\cos\left(2\tan^{-1}\frac{4}{3}\right)$, and then evaluate $\left(-\frac{375}{7}P+2\right)$.

(10-2-20-17)

The wrong thing you can do is to try to cling to something that's gone, or to recreate it.
Johnette Napolitano

1. Solve the radical equation: $\sqrt{t+6}-\sqrt{t-1}=1$.

2. Solve the equation: $7+3\sqrt[5]{z-2}=10$.

3. Solve the trigonometric equation: $\sin x - \sin 2x = 0$. If $c°$ is the nonzero solution in the first quadrant, find $\frac{c°}{3}$.

4. Solve the equation: $\cos 2x = \sin x$. For the solution $t°$ in the first quadrant, find $\frac{t°}{2}$.

(10-3-20-15)

Doing what is right isn't the problem; it's knowing what is right.

Lyndon B. Johnson

1. Simplify: $-10 + 2\sqrt{-20} - (-20 + \sqrt{-80})$.

2. Simplify, write your answer in a + bi form and evaluate $(-\frac{a}{b})$:
   $(-5 + \sqrt{-16}) - (3 + \sqrt{-4})$.

3. If k° is the solution in the first quadrant for the equation
   $8 - 12 \sin^2 x = 4 \cos^2 x$, evaluate $(\frac{4}{9} k°)$.

4. Solve the equation in degrees: $\sin 2x - \cos x - 2 \sin x + 1 = 0$.
   Then find one half of the nonzero solution in the first quadrant.

   (10-4-20-15)

Perhaps the world's second worst crime is boredom. The first is being a bore.

Cecil Beaton

1. Simplify in the a + bi form and find $|a|$: $\frac{5-3i}{i} + 7 i^2$.

2. In problem #1, find $|b|$.

3. A right triangle has a 30° angle. Find the hypotenuse if the longer leg is $10\sqrt{3}$.

4. A right triangle has a 60° angle. Find the longer leg, if the hypotenuse is $10\sqrt{3}$.

   (10-5-20-15)

Today is yesterday's effect and tomorrow's cause.

Philip Gribble

1. Simplify to the a + bi form and evaluate (- 4b): $\frac{3-2i}{1+i}$.

2. In problem #1, evaluate (12a).

3. Two of the angles of a triangle are 60° and 45° with their respective opposite sides c and b. Find the exact value of b if
   $c = 10\sqrt{6}$.

4. If t is the third side of the triangle from problem #3, evaluate
   $(t + 7 - 10\sqrt{3})$.

(10-6-20-17)

*Every man has his own courage, and is betrayed because he seeks in
himself the courage of other persons.*

Ralph Waldo Emerson

1. Solve by the square root property, then evaluate (10 + S) where
   S is the sum of the solutions of the equation: $5t^2 - 12 = 0$.

2. If m and n are the solutions of the equation: $-3x^2 + 21 = -6$,
   evaluate $(|m| + |n| + 1)$.

3. The two adjacent sides of the 60° angle of a triangle are 4 and 5.
   If P is the perimeter of the triangle, evaluate $(P + 11 - \sqrt{21})$.

4. The sides of a triangle are $5\sqrt{3}$, 10, and 5. Find half of the smallest angle of the triangle.

(10-7-20-15)

There is nothing to be gained by waiting for a better situation. You see where you are and you do what you can with that.

Jacob K. Javits

1. Solve the equation by completing the square, and then find the sum of the solutions: $x^2 = 10x - 16$.

2. Solve by completing the square, and find the absolute value of the product of the solutions: $3x^2 + 6x = 24$.

3. The two adjacent sides of the $45°$ angle of a triangle are 12 and 8. If A is the exact area of the triangle, evaluate $\left(\frac{5\sqrt{2}}{12} A\right)$.

4. The two shorter sides of a triangle with a $120°$ angle are 5 and 6. If A is the exact area of the triangle, evaluate $\left(\frac{2\sqrt{3}}{3} A + 2\right)$.

(10-8-20-17)

Failure is only postponed success as long as courage 'coaches' ambition. The habit of persistence is the habit of victory.

Herbert Kaufman

1. Solve by completing the square, and then find (5t) if t is the positive solution: $3a^2 - 4a = 4$.

2. Solve by completing the square. Then evaluate $(3\sqrt{6}\,|b|)$, where b is from the $(a + bi)$ form of the solutions.

3. An isosceles triangle with an angle of $120°$ has an area of $400\sqrt{3}$. Find the height of the triangle corresponding to its longest side.

4. The area of an equilateral triangle is $75\sqrt{3}$. Find its height.

(10-9-20-15)

Self-pity is our worst enemy and if we yield to it, we can never do anything wise in this world.

Helen Keller

1. Graph the parabola $f(x) = (2x - 1)^2 - 15$, then find $f(3)$.

2. Graph the parabola $g(x) = (x - 4)^2 - 12$, and find $x > 0$ such that $g(x) = 24$.

3. If y is the y-rectangular coordinate of a point given in polar form as $(-4, -\frac{\pi}{4})$, evaluate $(5y\sqrt{2})$.

4. For the point in polar coordinates $(-2, \frac{4\pi}{3})$ find its x-rectangular coordinate, t. Then evaluate $(2 + 15t)$.

(10-10-20-17)

This time, like all times, is a very good one if we but know what to do with it.

Ralph Waldo Emerson

1. Find the maximum of the function: $f(t) = -2t^2 - 4t + 8$.

2. Graph the parabola, then find its y-intercept: $g(z) = 3(z - 1)^2 + 8$.

3. For the point in rectangular coordinates $(10\sqrt{3}, 10)$ find its polar coordinates in the first quadrant (r, θ). What is the value of r?

4. In problem #3, find half of the angle θ in degrees.

(10-11-20-15)

Don't take anyone else's definition of success as your own.
(This is easier said than done.)

Jacqueline Briskin

1. For the parabola $f(x) = -2(x + 1)^2 + 50$, find the length of the interval on the x-axis where $f(x) \geq 0$.

2. Graph the function from problem #1, and find $\frac{1}{4}$ of its y-intercept.

3. Convert the equation $2x^2 + 2y^2 = 5y$ from rectangular coordinates to polar form, (r, θ). Then, if R is the nonzero value of r when $\theta = \frac{\pi}{6}$, evaluate $(16R)$.

4. For $\theta = \frac{\pi}{3}$ in problem # 3, find the nonzero value of r, and evaluate $(2 + 4r\sqrt{3})$.

(10-12-20-17)

First say to yourself what you would be, and then do what you have to do.

Epictetus

1. A parabola has the x-intercepts $\frac{(7 \pm \sqrt{29})}{2}$ and a minimum of - 4.5. Find 4 times the sum of the coordinates of the vertex.

2. If k is the x-coordinate of the vertex of the parabola in problem #1, evaluate $(2k + 6)$.

3. If M is the maximum of the function given in polar form: $r \sin \theta(1 - r \sin \theta) = 4 - r^2$, find $(5M)$.

4. Convert from polar to rectangular form, graph, and then find x when y = 3: $r^2 \cos^2 \theta - 6 = r^2 - r \cos \theta$.

(10-13-20-15)

Happiness comes of the capacity to feel deeply, to enjoy simply, to think freely, to risk life, to be needed.

Storm Jameson

1. Solve the equation by the quadratic formula: $2x^2 - 23x + 30 = 0$. What is the largest solution?

2. Solve the equation by the quadratic formula and then find the product of the solutions: $3x^2 - 27x = -42$.

3. Convert the complex number $(-\sqrt{3} + i)$ to polar form (with θ in degrees) and evaluate $\left(\frac{r\theta}{15}\right)$.

4. In problem #3, evaluate $\left(\frac{\theta}{5r} + 2\right)$.

(10-14-20-17)

You decide what it is you want to accomplish and then you lay out your plans to get there, and then you just do it. It's pretty straightforward.

Nancy Ditz

1. If P is the product of the solutions of the equation:
$$\frac{1}{2}x^2 - 0.4x + 1 = 0, \qquad \text{evaluate (5p).}$$

2. Find (3t), when t is the positive solution of the equation:
$$\frac{1}{3}x^2 = x + \frac{10}{3}.$$

3. For the complex numbers in trigonometric form:
$z = \cos 210° + i \sin 210°$ and $w = 2(\cos 190° + i \sin 190°)$, find $\frac{z}{w}$ and indicate the value of θ.

4. Simplify $\left[2\left(\cos\frac{5\pi}{16} + i\sin\frac{5\pi}{16}\right)\right]^4$, find θ in degrees, and then evaluate $\left(2 + \frac{\theta}{15}\right)$.

(10-15-20-17)

Where the willingness is great, the difficulties cannot be great.

Niccolo Machiavelli

1. Find the largest solution of the equation: $2t = 17 + \frac{30}{t}$.

2. If S is the sum of the solutions of the equation $5m - 4 = \frac{5}{m}$, find (20S).

3. Find all fourth roots of (- 16). If θ, *in degrees,* is the angle of the second fourth root, evaluate $\left(\frac{4\theta}{27}\right)$.

4. In problem #3, if α is the angle of the fourth root (in degrees), evaluate $\left(\frac{\alpha}{21} + 2\right)$.

(10-16-20-17)

Human behavior flows from three main sources: desire, emotion, and knowledge.

Plato

1. The area of a rectangle is 60. Find the product of 2 and the shorter side if one side is 2 more than twice the other.

2. In problem #1, find half of the perimeter of the rectangle.

3. If D is the length of the domain of the ellipse, evaluate (5D):
$$9(x - 2)^2 + 4(y + 1)^2 = 36.$$

4. In problem #3, if R is the length of the range of the ellipse, evaluate (2.5R + 2).

(10-17-20-17)

Discontent is the first step in the progress of a man or a nation.
Oscar Wilde

1. If they work individually, Sicha takes six more hours than her sister Micha to paint a room. If they work together, they take four hours to paint the same room. If S is the number of hours Sicha takes to paint the room alone, find $(S - 2)$.

2. In problem #1, find $(3M)$, where M is the number of hours Micha takes to paint the room alone.

3. Graph the ellipse: $x^2 + 4y^2 + 2x - 24y + 33 = 0$. If M is the y-coordinate of its highest point, evaluate $(5M)$.

4. In problem #3, let L be the smallest x-coordinate of the ellipse. Evaluate $|5L|$.

(10-18-20-15)

He that will have a perfect brother must resign himself to remaining brotherless.
Italian proverb

1. If $I = i^{127}$, evaluate $(10iI)$.

2. If $f(x) = x^2 + x + 18$, find $f(1 + i)$. Then write your answer in a + bi form and indicate the value of 'a'.

3. If L is the length of the major axis of the ellipse, evaluate $(5L)$:
$$(x + 1)^2 + 4y^2 = 24y - 32.$$

4. K is the y-coordinate of the center of the ellipse in problem #3. Find $(5K + 2)$.

(10-19-20-17)

There is no such thing as something for nothing.

Napoleon Hill

1. For the parabola $f(x) = 3x^2 - 12x + 8$, evaluate $(h - 2k)$, where h and k are respectively the x- and the y-coordinates of its vertex.

2. For $f(x)$ from problem #1, graph $g(x) = -f(x)$, and evaluate $(5M)$ where M is the maximum of $g(x)$.

3. Graph the hyperbola by locating the center, vertices, and the as-ymptotes: $x^2 - y^2 - 2x - 2y = 1$. If T is the larger of the two x-coordinates of the vertices, find $(10T)$.

4. In problem #3, if $(17, b)$ is a point on the asymptote with posi-tive slope, find b.

(10-20-20-15)

The most beautiful discovery true friends make is that they can go sepa-rately without growing apart.

Elizabeth Foley

1. Simplify: $\dfrac{(10\sqrt{3}+15)}{(\sqrt{12}-\sqrt{3})}(4 - 2\sqrt{3})$.

2. Find the vertex $V(h, k)$ of the parabola $g(x) = -2x^2 + 12x - 11$, sketch its graph, and evaluate $(h\cdot k)$.

3. Find the conic in rectangular coordinates represented in polar form by the equation $r = \dfrac{6}{2-\sin\theta}$. Then evaluate $(10b)$ if b is the y-coordinate of the center of the conic.

4. In problem #3, if the point of rectangular coordinates $(2\sqrt{3}, k)$ is on the graph of the conic, evaluate $(2 + 7.5k)$.

(10-21-20-17)

Friendship makes prosperity more brilliant, and lightens adversity by dividing and sharing it.

Cicero

1. A water pipe runs diagonally in a rectangular garden of area 48ft². If one side of the garden is 4ft less than twice the other side, find the length of the pipe.

2. For the garden in problem #1, if one of the shorter sides of the garden is against a house, how many feet of fencing is needed to close in the garden?

3. Find the rectangular equation of the curve given in parametric form: $x = t^{1.5}$ and $y = 2t + 4$, $t \geq 0$. Then find b if the point $(16\sqrt{2}, b)$ is on that curve.

4. The point of rectangular coordinates (14, k) is on the curve given in parametric form: $x = 2t^2 + 6$ and $y = 5 - t$. Evaluate $(5k + 2)$.

(10-22-20-17)

Only a life lived for others is a life worth-while.

Albert Einstein

1. Find the shorter side of a rectangular flower garden of maximum area enclosed with 40ft of fencing.

2. If A is the maximum area in problem #1, evaluate $(.25A - 2)$.

3. Given that (0, T) are the rectangular coordinates of a point on the curve represented in parametric form $x = 3\sin t$, $y = 4\cos t + 2$, $0 \leq t \leq \frac{\pi}{2}$, evaluate $\left(\frac{10}{3}T\right)$.

4. Find the rectangular equation represented in parametric form $x = 4t - 2$ and $y = 1 - t$. Then find y when $x = -58$.

(10-23-20-15)

Do things for others and you'll find your self-consciousness evaporating like morning dew.

Dale Carnegie

1. The base of a triangular piece of land of area 120 square yards is 6 yards less than triple its height. Find the height.

2. Find the base of the triangular piece of land from problem #1.

3. Solve the system for x: $x + y - 2z = -3$
   $2x + 2y - z = 39$
   $-x + 3y - 3z = -44$

4. Solve the system from problem #3 for z.

(10-24-20-15)

The deepest principle in human nature is the craving to be appreciated.

William James

1. In an arithmetic sequence $a_1 = 1$ and d = 3. Find a_4.

2. An arithmetic sequence has $a_3 = 10$ and $a_8 = 40$. Find $a_5 + 3$.

3. Solve the system for x: $\frac{1}{2}x + \frac{2}{3}y = 20$
   $\frac{3}{5}x + \frac{1}{3}y = -7$

4. Solve the system from problem #3 for y.

(10-25-20-15)

Who you are is a necessary step to being who you will be.

Emmanuel

1. If $a_n = 3n - 2$, list the first 6 terms of the sequence and identify a_4.

2. If in an arithmetic sequence $a_4 = 8$ and $a_{10} = 20$, find a_{13}.

3. If matrix A = $\begin{bmatrix} 2 & -5 \\ 3 & 1 \end{bmatrix}$ and B = $\begin{bmatrix} 1 & b \\ -4 & 2 \end{bmatrix}$, find b such that A + B = $\begin{bmatrix} 3 & 15 \\ -1 & 3 \end{bmatrix}$.

4. In problem #3, find b if B – A = $\begin{bmatrix} -1 & 20 \\ -7 & 1 \end{bmatrix}$.

(10-26-20-15)

Your problem is you're ... too busy holding onto your unworthiness.

Ram Dass

1. Evaluate the sum: $\sum_{k=1}^{4} k$.

2. In a geometric sequence $a_2 = 9$ and $a_4 = 81$. Find a_3 for r > 0.

3. If matrix A = $\begin{bmatrix} 10 & x & -5 \\ 5 & -20 & y \end{bmatrix}$, find x if $\frac{3}{5}A = \begin{bmatrix} 6 & 12 & -3 \\ 3 & -12 & 9 \end{bmatrix}$.

4. Under the conditions in problem #3, find (y + 2).

(10-27-20-17)

We can see well into the past; we can guess shrewdly into the future; but that which is rolled up and muffled in impenetrable folds is today.

Ralph Waldo Emerson

1. Evaluate: $\frac{5!}{3!\cdot 2}$.

2. Evaluate (n! + n) for n = 4.

3. For A = $\begin{bmatrix} 2 & -1 \\ 3 & x \end{bmatrix}$, find x if $A^{-1} = \begin{bmatrix} \frac{20}{43} & \frac{1}{43} \\ \frac{-3}{43} & \frac{2}{43} \end{bmatrix}$.

4. If M = $\begin{bmatrix} 1 & -2 \\ -1 & y \end{bmatrix}$, find y such that $M^{-1} = \frac{1}{13}\begin{bmatrix} 15 & 2 \\ 1 & 1 \end{bmatrix}$.

(10-28-20-15)

Everyman's life lies within the present, for the past is spent and done with, and the future is uncertain.

Marcus Aurelius

1. Simplify: $-\sum_{k=1}^{5}(1-k)$.

2. Simplify: $-1 + \sum_{k=1}^{5}(1-k)^2$.

3. Use matrices to find the value of y: x − y + z = - 4
   2x + 2y − z = 27
   -x + y + 2z = 49.

4. In problem #3, find z.

(10-29-20-15)

One's feelings waste themselves in words; they ought all to be distilled into action which bring results.

Florence Nightingale

1. In an arithmetic sequence $a_3 = 24$ and $a_7 = 64$. Find the common difference, d.

2. For the sequence in problem #1, evaluate $(a_{k+3} - a_k)$ for $\forall\, k \geq 1$.

3. Find the value of the determinant $D = \begin{vmatrix} 1 & 2 & 3 \\ -1 & 1 & 0 \\ -2 & 1 & 2 \end{vmatrix}$. Then evaluate $(11 + D)$.

4. If the value of the determinant $\begin{vmatrix} -1 & 1 & -2 \\ x & 1 & -1 \\ 2 & 0 & -3 \end{vmatrix}$ is 50, find x.

(10-30-20-15)

Before everything else, getting ready is the secret of success.

Henry Ford

1. In the geometric sequence with $a_1 = 1$ and $r = 2$, find $(a_3 + a_4 - 2)$.

2. In the sequence from problem #1, evaluate $(a_{17} - 1)$.

3. Find x from the determinant $D = \begin{vmatrix} 1 & -1 & x \\ -1 & 0 & 2 \\ 3 & 4 & -1 \end{vmatrix}$ if the value of D is (- 93).

4. If x = - 1 in determinant D from problem #3, find the value of the determinant D, and then evaluate $(-2D - 1)$.

(10-31-20-17)

If one advances confidently in the direction of his dreams, and endeavors to live the life he has imaged, he will meet with success unexpected in common hours.

Henry David Thoreau

1. If a_2 = 13 and a_5 = 46 in an arithmetic sequence, find the common difference, d.

2. In the sequence from problem #1, evaluate $\frac{(2a_4 - a_7)}{2}$.

3. Use Cramer's Rule to solve for x in the system:
$$2x - 3y = -5$$
$$-x + 2y = 10$$

4. In problem #3, use Cramer's Rule to solve for y.

(11-1-20-15)

To think is easy. To act is difficult. To act as one thinks is the most difficult of all.

Johann von Goethe

1. In the arithmetic sequence with the first term 1 and the common difference 2, find the sixth term.

2. In the sequence from problem #1, evaluate $\frac{(a_6 - a_4)}{(a_{22} - a_{21})}$.

3. Use Cramer's Rule to solve for y in the system:
$$x + y - z = 8$$
$$3x + 2y - 3z = 4$$
$$x - y + z = -2$$

4. Use Cramer's Rule to solve for z in problem #3.

(11-2-20-15)

The reward of a thing well done is to have done it.
Ralph Waldo Emerson

1. If a_n = 3n and a_k = 33, find k.

2. If the first and the seventh terms of an arithmetic sequence are respectively – 9 and 9, find the common difference.

3. Write the partial fraction decomposition of the rational expression: $\frac{5x-20}{x\,(x-1)}$. What is the positive value of the larger numerator?

4. In problem #3, find the positive value of the smaller numerator.

(11-3-20-15)

Out of every fruition of success, no matter what, comes forth something to make a new effort necessary.
Walt Whitman

1. In an arithmetic sequence, d = - 2 and a_1 = 50. If a_k = 30, find k.

2. In the sequence from problem #1, find a_{24}.

3. Find the partial fraction decomposition of the rational expression: $\frac{(35x^2 - 39x + 58)}{(x^2 + 4)(x-2)}$.
   Then indicate the value of m in the numerator of the (mx + b) form.

4. Find the constant numerator in problem #3.

(11-4-20-15)

Along with success comes a reputation for wisdom.

Euripides

1. A geometric sequence has $a_1 = 2$ and $r = 5$. If the sum of the first k terms of the sequence is $\frac{1}{2}[5(25)^5 - 1]$, find k.

2. In the sequence from problem #1, evaluate $\sqrt[3]{\frac{a_6}{a_3}}$.

3. Write the rational expression in partial fraction decomposition form, and identify the larger numerator when x = 7:
$$\frac{(8x^2 - 7x + 17)}{(x^3 - 2x^2 + 3x - 6)}.$$

4. In problem #3, find the smaller numerator, S, when x = 3. Then find (3S).

(11-5-20-15)

He is blessed over all mortals who loses no moment of the passing life.

Henry David Thoreau

1. For the geometric sequence 3, 6, 12, 24, ... find the sum of the first seven terms, S, and evaluate: $\frac{S-7}{34}$.

2. In the sequence from problem #1, evaluate $\frac{a_7}{32}$.

3. Solve the nonlinear system of equations and evaluate (x·y) when x ≠ 0:
$$2x - y = 3$$
$$y^2 = 4x + 9$$

4. In problem #3, evaluate (3y + 2) for x ≠ 0.

(11-6-20-17)

Great opportunities come to all, but many do not know they have met them. The only preparation to take advantage of them is ... to watch what each day brings.

Albert E. Dunning

1. For the arithmetic sequence 4, 8, 12, 16, ..., evaluate $\frac{(a_{14}-1)}{5}$.

2. In the sequence from problem #1, the sum of the first k terms is 112. Find k.

3. Solve the nonlinear system for y:
$$x^2 + 2x = -y^2 + y + 4$$
$$x + 2 = \frac{4-y}{x}$$
For the nonzero value of y, find (10y).

4. Solve the nonlinear system:
$$x - y = 2$$
$$x^2 - 9y = -2.$$
If (x,y) is a solution pair, evaluate (x·y + 2) for the largest value of (x + y).

(11-7-20-17)

Two persons cannot long be friends if they cannot forgive each other's little failings.

Jean de la Bruyere

1. In the geometric sequence 2, $\frac{2}{3}x$, $\frac{2}{9}x^2$, $\frac{2}{27}x^3$, ... find the common ration when x = 33.

2. In the sequence from problem #1, find the sum of the first five terms, S, and then evaluate $\sqrt{S+2}$ for x = 6.

3. Graph the system of inequalities in the first quadrant, then evaluate (4x), where x is the largest x-coordinate of the corner points of the solution:
$$3x - y \leq 6$$
$$x + y \leq 14$$

4. In problem #3, evaluate (2y − 1), where y is the second largest y-coordinate of the corner points of the solution.

(11-8-20-17)

I always felt that the great high privilege, relief and comfort of friend-ship was that one had to explain nothing.

Katherine Mansfield

1. In the geometric series $2x - 4x + 8x - 16x + 32x - ...$ find k if
$a_k = \dfrac{8^4}{2}x.$

2. How many terms of the series in problem #1 need to be added, for their sum to be $342x$?

3. Graph the system of nonlinear inequalities, and then evaluate $(4x)$, where x is the x-coordinate of the highest corner point of the solution:
$$-x^2 + 7 \le 6y$$
$$y \le -2x + 7$$

4. In problem #3, evaluate $(|5y| + 2)$, where y is the y-coordinate of the same corner point.

(11-9-20-17)

It is a good thing to be rich, it is a good thing to be strong, but it is a bet-ter thing to be beloved of many friends.

Euripides

1. Write $0.454545...$ in simplified fraction form. What is the de-nominator of the fraction?

2. Find the limit of the infinite geometric series:
$5 + 2.5 + 1.25 + 0.625 + ...$

3. Maximize $z = x + 2y$ when x and y are positive, under the condi-tions:
$$2x - y \le 4$$
$$y \le -4x + 10$$

4. Maximize $z = 4x + 3y$ in the first quadrant, under the conditions:
$$x + 3y \le 6$$
$$3x + 4y \le 13$$

(11-10-20-15)

Live now, believe me, wait not till tomorrow, gather the roses of life to-day.

Pierre de Ronsard

1. A ball is dropped from a height of 5.5m and each rebound is half of the previous one. Find half of the total distance traveled by the ball until it comes to rest.

2. If p and q are the fifth and respectively the third coefficients in the ordered expansion of $(a + b)^7$, evaluate $(p - q - 3)$.

3. Find b in the equation of the parabola $y = x^2 + bx + c$, such that its graph contains the points of coordinates (-1, -4) and (1, 36).

4. Find c in problem #3.

(11-11-20-15)

Sunshine is delicious, rain is refreshing, wind braces us up, snow is exhilarating; there is really no such thing as bad weather, only different kinds of good weather.

John Ruskin

1. The perimeter of a rectangular garden is 46ft. The area of the garden is 132ft^2. Find its shorter side.

2. Find the longer side of the garden in problem #1.

3. In the circle of equation $x^2 + y^2 + Ax + By + 5 = 0$, find the product of the coordinates of its center, if (4, 11) and (4, -1) are points on the circle.

4. In the circle from problem #3, evaluate (3R - 1), where R is the radius of the circle.

(11-12-20-17)

Whatever the ups and downs of detail within our limited experience, the larger whole is primarily beautiful.

Gregory Bateson

1. Two parallel lines are cut by a transversal at an acute angle. If the acute angle is x and its adjacent angle is (3x + 4), evaluate $\left(\frac{x}{4}\right)$.

2. In problem #1, find x if its adjacent angle is (10x +37).

3. If the points (1, -10), (-1, -40), and (2, -10) are on the graph of the parabola $y = ax^2 + bx + c$, find $|c|$.

4. In problem #3, evaluate (2 + b).

(11-13-20-17)

I make the most of all that comes and the least of all that goes.

Sara Teasdale

1. Suppose 7 million pounds of coffee are sold when the price is $6.5 per pound, and 10 million are sold at $5 per pound. At this rate, how many million pounds would be sold at $4.5 per pound?

2. In problem #1, how many million pounds of coffee will be sold at $3 per pound?

3. In the sequence $\{(-1)^{n+1}(2n + 3)\}$, find the ninth term, N, and evaluate N-1.

4. In the sequence from problem #3, find the absolute value of the sixth term.

(11-14-20-15)

Let us go singing as far as we go; the road will be less tedious.

Virgil

1. Twice a smaller number, plus a larger number is 37. Their difference is 4. Find the smaller number.

2. In problem #1, find the larger number.

3. In the sequence given by the general term $a_n = \frac{2^n}{n^2}$, evaluate $(5a_8)$.

4. In the sequence from problem #3, evaluate $[2 + 5(a_8 - a_4)]$.

(11-15-20-17)

What one has, one ought to use; and whatever he does, he should do with all his might.

Cicero

1. The area of a triangle is $88ft^2$. If the height of the triangle is 5ft shorter than the corresponding base, find the height.

2. Find the base in problem #1.

3. A sequence is given by: $a_1 = -3$ and $a_n = 4 + a_{n-1}$. Find $(a_{11} - a_6)$.

4. In the sequence from problem #3, find $\frac{1}{3}a_{13}$.

(11-16-20-15)

*Vigor is contagious, and whatever makes us either think or feel strongly
adds to our power and enlarges our field of action.*

Ralph Waldo Emerson

1. A store sells two kinds of notebooks: one kind for $2 and the other for $3 a piece. One day the store sold 28 notebooks and made $67. How many $3-notebooks were sold?

2. In problem #1, how many $2-notebooks were sold?

3. If a sequence is given by: $a_1 = 4$ and $a_n = -\frac{1}{4}a_{n-1}$, find $[256(a_5 - a_4)]$.

4. In the sequence given by: $a_1 = 2$ and $a_n = 3 - a_{n-1}$, find the sum of the first 10 terms.

(11-17-20-15)

*We are more often frightened than hurt; and we suffer more from imag-
ination than from reality.*

Marcus Annaeus Seneca

1. Greta works twice as fast as her brother Jay. Working together, they finish a job in 6 hours. If G is the number of hours Greta takes to do the same job alone, evaluate (G + 2).

2. In problem #1, how many hours does Jay take to do the same job alone?

3. In the sequence $\{2n^2 - 1\}$, evaluate $\frac{(a_6 - a_4)}{2}$.

4. In the sequence given by $a_n = 2n^2 - 2$, evaluate $[\frac{3}{4} \cdot \frac{(a_6 - a_4)}{2} + 2]$.

(11-18-20-17)

*If you let fear of consequence prevent you from following your deepest
instinct, then your life will be safe, expedient and thin.*
Katharine Butler Hathaway

1. If p and t are the solutions to the equation: $2(7 - x)^2 + 8 = 16$,
   evaluate $(p + t - 3)$.

2. The area of a rectangular garden is $84ft^2$. One side is 5ft shorter
   than the other. Find half of the perimeter of the garden.

3. For the sequence $\{2n^3\}$ find the value of $\left(\frac{1}{10}S\right)$, where S is the
   sum of the first four terms.

4. For the sequence in problem #3, evaluate $(\frac{1}{30}S + 2)$, where S is
   the sum of the first five terms.

(11-19-20-17)

I am not afraid of tomorrow, for I have seen yesterday and I love today.
William Allen White

1. If m and n are the solutions of the equation: $x + \frac{8}{x} = -9$, evaluate
   $(|m + n| + 2)$.

2. Two angles are supplementary. One is 40° smaller than the other.
   Find the complement of the smaller angle.

3. In the sequence 0, 4, 8, 12, ..., find $(a_{16} - a_{11})$.

4. If S is the sum of the first 30 terms in the sequence from problem
   #3, evaluate: $\left(2 + \frac{S}{116}\right)$.

(11-20-20-17)

180

To see what is right, and not do it, is want of courage.

Confucius

1. One week, James tutored 32 students and made a total of $258. If he charged $12 for private tutoring and $6 for group tutoring, how many students were tutored privately?

2. In problem #1, how many students were tutored in group?

3. For the sequence $\{(-1)^n \cdot 2^{3-n}\}$, evaluate $(-7S - 1)$, where S is the sum of the first 3 terms.

4. For the infinite series based on the sequence from problem #3, evaluate: $(-\frac{45}{8}S + 2)$, where S is its sum.

(11-21-20-17)

Keep away from people who try to belittle your ambitions. Small people always do that, but the really great make you feel that you, too, can become great.

Mark Twain

1. In a trapezoid of height 6, one of the two parallel sides is half of the other. If the area of the trapezoid is 99, find the shorter of the parallel sides.

2. In problem #1, if m is the length of the median of the trapezoid and b is the shorter parallel side, evaluate: $2m - b$.

3. Find: $\frac{1}{3}\sum_{k=1}^{4}(-3)^k$.

4. Find: $\sum_{k=1}^{5}(3k - 6) + 2$.

(11-22-20-17)

There is a time for departure, even when there is no certain place to go.

Tennessee Williams

1. Maddie paid a 22% tip for a meal. If her total bill was $61, what was the amount of the tip?

2. The largest side of a triangle is twice the smallest. The smallest is 3 shorter than the middle side, while the largest is 2 longer than the middle side. Find the perimeter of the triangle.

3. Use the Principle of Mathematical Induction to verify the identity: $S_n = \sum_{k=1}^{n} 2^{k-1} = 2^n - 1$. Then evaluate $(S_5 - 11)$.

4. In problem #3, evaluate $(272 - S_8)$.

(11-23-20-17)

Prayer indeed is good, but while calling on the gods, a man should himself lend a hand.

Hippocrates

1. Solve the equation: $1 + \sqrt{7x + 23} = x$.

2. Solve the equation: $3 - (2x - 21)^{\frac{1}{3}} = 0$.

3. Find the coefficient of the fourth term in the ordered binomial expansion of $(a + b)^6$.

4. Calculate $\frac{C}{14}$, where C is the coefficient of the seventh term in the ordered binomial expansion of $(x + y)^{10}$.

(11-24-20-15)

Be glad of life because it gives you the chance to love, and to work, and to play, and to look up at the stars.

Henry Van Dyke

1. Find the sum of the solutions of the equation: $2 + \dfrac{5}{x^2} = \dfrac{22}{x}$.

2. Two positive integer numbers differ by 9. One number is one more than double the other. Find the sum of the numbers.

3. Find $\dfrac{T}{56}$ if T is the coefficient of x^4 in the expansion of $(1- 2x)^8$.

4. Find the coefficient of x in the expansion of $(3x - 1)^5$.

(11-25-20-15)

You give but little when you give of your possessions. It is when you give of yourself that you truly give.

Kahlil Gibran

1. A right triangle has the sides $(x - 5)$, $(x + 2)$, and $(x + 3)$. Evaluate $(x + 1)$.

2. For the parabola $f(x) = -\dfrac{1}{3}x^2 + 4x - 7$, evaluate $(h \cdot k - 4)$, if the point (h, k) is its vertex.

3. If P is the number of permutations of 3 elements chosen from a set of 6, find $(\dfrac{1}{6}P)$.

4. Calculate the number of combinations of 4 objects you can make from a set of 6.

(11-26-20-15)

Goodness is the only investment that never fails.

Henry David Thoreau

1. Simplify the radical expression $2\sqrt[4]{x^2} - \dfrac{\sqrt[4]{x^6}}{x}$, and then evaluate it for x = 121.

2. Simplify the expression $\dfrac{9}{i^{91}}$, write the answer in (a + bi) form, and then   evaluate (3b).

3. If n is the number of ways 5 people can be lined up in a row, evaluate $\dfrac{n}{6}$.

4. Evaluate (C − 18), where C is the number of different committees of 3 people that can be formed from a group of 7.

(11-27-20-17)

To throw away an honest friend is, as it were, to throw your life away.

Sophocles

1. If $E = \dfrac{2}{\sqrt[3]{2}} \cdot \dfrac{\sqrt[3]{16}}{2^{-2}}$, find (E − 5).

2. For the expression in problem #1, evaluate: $49^{\frac{1}{2}}\sqrt{E}$.

3. If P is the probability of two heads to come up when you toss two fair coins, evaluate (80P).

4. The six faces of a fair die are numbered from 1 to 6. If P is the probability of getting a multiple of three at one roll of the die, calculate (45P + 2).

(11-28-20-17)

All the troubles of life come upon us because we refuse to sit quietly for a while each day in our rooms.

Blaise Pascal

1. In the formula $V = \sqrt{\dfrac{2U}{C}}$, solve for C, and then evaluate C when U = 22 and V = 2.

2. If d is the length of the diagonal of the rectangular flower garden of maximum area you can enclose with 16ft of fencing, evaluate $(d^2 - 3)$.

3. Four fair coins are tossed at the same time. If P is the probability of obtaining at least three heads, evaluate (50P).

4. In problem #3, if P is the probability of obtaining one, two, or three heads, find (25P + 2).

(11-29-20-17)

If we go down into ourselves, we find that we possess exactly what we desire.

Simone Weil

1. For $f(x) = x^2 - 3x$, find f(1 + i) in the (a + bi) form, and then evaluate $(a^2 + 2b^2)$.

2. Use complex numbers in the (a + bi) form to factor: $25 + 36t^2$. Then, evaluate $|ab|$ for either factor.

3. An urn contains three red balls and four white balls. A random ball is extracted twice, without being placed back in the urn. If P is the probability of extracting a white ball the second time, evaluate (35P).

4. Solve problem #3, if P is the probability of extracting a red ball the second time.

(11-30-20-15)

Austere perseverance, harsh and continuous, may be employed by the least of us and rarely fails of its purpose, for its silent power grows irresistibly greater with time.

Johann von Goethe

1. Simplify the expression: $\sqrt[3]{54x - 27}$. Then evaluate your answer for x = 32.5.

2. Multiply and simplify: $(\sqrt[3]{16} + \sqrt[3]{12} + \sqrt[3]{9})(\sqrt[3]{4} - \sqrt[3]{3})$.

3. Find the mean of the following set of data: {23, 18, 7, 28, 12, 32}.

4. Find the median of the following set of data: {21, 17, 9, 6, 13, 33}.

(12-1-20-15)

Geometry (1, 2) and Calculus (3, 4)

Bad times have a scientific value. These are occasions a good learner would not miss.

Ralph Waldo Emerson

1. Two sides of a right triangle are consecutive numbers. The third side is the shortest and is equal to 5. Find the longer leg of the triangle.

2. In problem #1, if D is the difference between the hypotenuse and the shorter leg, find $\sqrt[3]{D}$.

3. Find: $\lim\limits_{x \to 2} \dfrac{(x^2 - x + 58)}{(x^2 - 1)}$.

4. Evaluate L = $\lim\limits_{x \to -2} \dfrac{(x^2 - x - 6)}{(x^3 + 8)}$. Then find (- 36L + 2).

(12-2-20-17)

Fear is only an illusion. It is the illusion that creates the feeling of separateness – the false sense of isolation that exists only in your imagination.

Jeraldine Saunders

1. The shortest distance between point P and line l is PB = x. Point A is on line l such that AB = $3\sqrt{15}$. Knowing that PA = 4PB, find the length of PA.

2. In problem #1, evaluate: $\frac{(PB+PA+AB)}{(5+\sqrt{15})}$.

3. Find: L = $\lim_{x\to\infty} \frac{\sqrt{(4x^2-9)}}{x-5}$. Then evaluate (10L).

4. Find: L = $\lim_{x\to\infty}\left(\sqrt{x^2+6x+3} - x\right)$, then evaluate (5L + 2).

(12-3-20-17)

Time is a great manager: it arranges things well.

Pierre Corneille

1. M is a point on a line tangent at point P to a circle of center O. Find MP if   MO = $4\sqrt{10}$ and the area of the circle is 16π.

2. In problem #1, if A is the area of triangle MOP, evaluate $\left(\frac{A}{6}\right)$.

3. Evaluate the limit: L = $\lim_{x\to\frac{\pi}{2}} \frac{4\sin x}{1-5\cos x}$. Then find (5L).

4. Find: L = $\lim_{x\to2}\left(\frac{1}{x-2} - \frac{5}{x^2+x-6}\right)$. Then evaluate (2 + 75L).

(12-4-20-17)

Success follows doing what you want to do. There is no other way to be successful.

Malcolm Forbes

1. A right triangle has its sides proportional under the same ratio to three consecutive numbers. If the hypotenuse of the given triangle is 15, find the perimeter of a triangle formed with the three consecutive numbers.

2. What is the largest side of the triangle formed with the consecutive numbers in problem #1?

3. Use the definition of the derivative to evaluate $4f'(2)$, if $f(x) = x^2 + x$.

4. For the function in problem #3, find the equation of the tangent line to f(x) at the point (- 1, 0). Then find (- 15b + 2), where b is the y-intercept of the tangent line.

(12-5-20-17)

To finish the moment, to find the journey's end in every step of the road, to live the greatest number of good hours, is wisdom.

Ralph Waldo Emerson

1. Evaluate half of the area of a right triangle whose sides are consecutive even integers.

2. Find one fourth of the perimeter of the triangle in problem #1.

3. For the function $f(x) = (2x^2 - x + 3)^2$, find $\frac{5}{6}f'(1)$.

4. If $g(x) = \frac{2x-1}{\sqrt{x+4}}$, evaluate $(2 + 30g'(5))$.

(12-6-20-17)

You always get negative reactions. If you worry about that, you would never do anything.

Tom Monaghan

1. Segment AB crosses line I at C, under a 60° angle. Two triangles are formed by dropping perpendiculars from A and B to line I at D and E respectively. If the perimeters of triangles ADC and BEC are respectively $(21 + 7\sqrt{3})$ and $(36 + 12\sqrt{3})$, find the shortest side of triangle BEC.

2. In problem #1, find the shortest side of triangle ADC.

3. If $h(t) = e^t(t^2 + t + 1)$, find $h'(t)$, then evaluate $10h'(0)$.

4. Let $g(x) = \ln(x^3 e^x) + 13x$. Find $g'(x)$ and then evaluate $[2 + g'(3)]$.

(12-7-20-17)

Our doubts are traitors, and make us lose the good we often might win, by fearing to attempt.

William Shakespeare

1. The apothem of a regular hexagon is $6\sqrt{3}$. Find the radius of the circle circumscribed to the hexagon.

2. If a right triangle in inscribed in a circle of radius 5, and its legs are consecutive even numbers, find the longer leg.

3. For the function $f(x) = e^{\sin x} + \sin(e^x) - \sin(x + 1)$, find $f'(x)$. Then evaluate $(f'(0) + 19)$.

4. If $g(t) = \tan(\cos^{-1}t)$, find $g'(t)$. Then evaluate: $[-\frac{45}{8}g'\left(\frac{\sqrt{3}}{2}\right) + 2]$.

(12-8-20-17)

Freedom is nothing else but a chance to be better, whereas enslavement is a certainty of the worst.

Albert Camus

1. In triangle ABC, DE is parallel to BC, with D on AB and E on AC. If BC is 3 units longer than DE, and AE is 15 when EC is 5, find BC.

2. In problem #1, find DE.

3. Find an equation of the tangent line to the curve $x^2 - 4xy + y^2 = 13$ at point (1, -2). If m is the slope of this tangent line, evaluate: 16m.

4. For the curve in problem #3, if b is the y-intercept of the tangent line at point (1, 6), evaluate: $\frac{60}{13}b + 2$.

(12-9-20-17)

Doubts and mistrust are the mere panic of timid imagination, which the steadfast heart will conquer, and the large mind transcend.

Helen Keller

1. A square is inscribed in a circle of area 72π. Find the side of the square.

2. Point P is outside a circle, and PA and PB are two tangents to the circle, where A and B are corner points of the square inscribed in the circle. If PA = 5, find the diameter of the circle.

3. If t is the x-coordinate of the point where the tangent line to the function $f(x) = [\ln(x - 4)]^2$ is horizontal, evaluate (4t).

4. Obtain the double angle formula for the sin function by differentiating $\cos 2x = \cos^2 x - \sin^2 x$. Use these two formulas to find T = tan 2x when $x = \frac{\pi}{6}$, and then evaluate $(2 + 5T\sqrt{3})$.

(12-10-20-17)

It is not because things are difficult that we do not dare, it is because we do not dare that they are difficult.

Marcus Annaeus Seneca

1. Two angles of a triangle are complementary. If the diameter of the circle circumscribed to the triangle is 13 and the shortest side of the triangle is 5, find the third side of the triangle.

2. The sides of a triangle are consecutive odd numbers. If the difference between the squares of the two larger sides is 56, find the shortest side of the triangle.

3. Find the local extreme, T, of the function $f(x) = 2x^3 - 3x^2 - 36x + 6$ in the interval [- 3, - 1]. Then evaluate $(T - 30)$.

4. In problem #3, find the local extreme, M, in the interval [2, 4], then evaluate $(- 0.2M + 2)$.

(12-11-20-17)

Only those who dare to fail greatly can ever achieve greatly.

Robert F. Kennedy

1. The perimeter of an equilateral triangle is $24\sqrt{3}$. Find the height of the triangle.

2. One of the equal sides of an isosceles triangle is 10, and the height dropped to that side is 9.6. Find the unequal side of the triangle if the other height is 8.

3. For the function $g(x) = x + \cos 2x$, find the x-coordinate, t, in degrees, of the local extreme point in the interval [0, 90°]. Then evaluate $(t + 5)$.

4. If b is the y-coordinate of the extreme point in problem #3, evaluate $(2 + b - \frac{\sqrt{3}}{2})$.

(12-12-20-17)

The ambitious climb high and perilous stairs, and never care how to come down, the desire of rising hath swallowed up their fear of a fall.

Thomas Adams

1. The two parallel sides of a trapezoid are consecutive integers. Find the shorter of the two parallel sides if the area of the trapezoid is 100 and the height is 8.

2. Find the longer of the two parallel sides of the trapezoid in problem #1.

3. For the function $f(x) = x^4 - 24x^2 - 25$, find the absolute minimum, M, and then evaluate $(189 + M)$.

4. For the function in problem #3, find an inflection point, (x, I), and then evaluate $(-\frac{1}{7}I + 2)$.

(12-13-20-17)

The method of the enterprising is to plan with audacity and execute with vigor.

Christian Bovee

1. The two parallel sides of a trapezoid are consecutive even numbers. If the median of the trapezoid is 13, find the shorter of the two parallel sides.

2. In problem #1, find the longer of the two parallel sides of the trapezoid.

3. If X is the positive x-coordinate of an inflection point on the graph of $f(x) = 2x + \ln(x^2 + 2)$, evaluate $(10X\sqrt{2})$.

4. If x is the angle in degrees in the first quadrant for which $f(x) = \cos^2 x - 2\sin x$ has an inflection point, evaluate $(0.5x + 2)$.

(12-14-20-17)

Those who are the most persistent, and work in the true spirit, will invariably be the most successful.

Samuel Smiles

1. A square is inscribed in a circle and the circle is inscribed in a larger square. If the area of the smaller square is 72, find the side of the larger square.
2. The hypotenuse of a right triangle is 25 and the area of the triangle is 150. Find the shorter leg if it is a multiple of 5.
3. Find the point (x, y) on the line y − x = 3, that is closest to the point (4, 0) (hint: minimize the distance using derivatives). Then evaluate (40x).
4. Use derivatives to find the closest point on the curve $x^2 + y^2 = 9$ to the point (4, 3). If t is the x-coordinate of that point, evaluate $(2 + \frac{25}{4}t)$.

(12-15-20-17)

Time isn't a commodity, something you pass around like cake. Time is the substance of life. When anyone asks you to give your time, they're really asking for a chunk of your life.

Antoinette Bosco

1. The three angles of a triangle are consecutive multiples of 30. If the middle side of the triangle is $4\sqrt{3}$, find the sum of the other two sides of the triangle.
2. The area of an equilateral triangle is $192\sqrt{3}$. Find the radius of the circle circumscribed to the triangle.
3. Use Riemann sums to find the area, A, bounded by the graph of the function $f(x) = x^2 - x$ and the x-axis for x ∈ [1, 2]. Then evaluate (24A).
4. Use integrals to find the area, A, in problem #3, and then find (18A + 2).

(12-16-20-17)

Courage is the most important of all virtues, because without it we can't practice any other virtue with consistency.

Maya Angelou

1. Point P is outside a circle of center O. Line PO intercepts the circle at A and B, with A closer to P. AE and BC are perpendiculars to the tangent line from P to the circle at D, with E and C on the tangent line. If angle APE is 30° and DC = $3\sqrt{3}$, find the diameter of the circle.

2. In problem #1, evaluate (PB − 1).

3. Evaluate the integral: I = $\int_0^1 x(x^2 + 1)^5 dx$. Then find $\left(\frac{80}{21} I\right)$.

4. Find the family of functions f(x) = $\int \frac{x-2}{\sqrt{x^2-4x}} dx$. If the point $(5, 15 + \sqrt{5})$ is on the graph of f(x), find $[2 + f(4) + \frac{(2\sqrt{5})}{3}]$.

(12-17-20-17)

It is better by noble boldness to run the risk of being subject to half of the evils we anticipate than to remain in cowardly listlessness for fear of what might happen.

Herodotus

1. The tangents from point P to a circle of center O, intercept the circle at A and B, with angle POB = 45°. If the area of the circle is 72π, find OP.

2. Find $\frac{1}{4}$ of the area of APBO in problem #1.

3. Find g(x) = $\int \frac{\cos x}{1+\sin x} dx$. If the point $(\frac{\pi}{2}, 20 + \ln 2)$ is on the graph of g(x), find g(0).

4. Find the area, A, between f(x) = e^x and g(x) = x^2 for x ∈ [0, 1]. Then evaluate: $(\frac{45A}{3e-4} + 2)$.

(12-18-20-17)

He who loses wealth loses much; he who loses a friend loses more;
but he who loses courage loses all.

Miguel de Cervantes

1. Two parallel lines are cut by a transversal at points A and B at an angle of 60°. If AB = $8\sqrt{3}$, find the distance between the two parallel lines.

2. The apothem of a regular hexagon is $3\sqrt{3}$. Find the diameter, D, of the circle circumscribed to the hexagon, and evaluate (2D – 5).

3. If A = $\int_1^{e^\pi} \frac{\sin(\ln x)}{x}$ dx, evaluate (10A).

4. Find the area, S, bounded by x – y = 0 and x = - y² + 3y. Then evaluate: $\left(\frac{45}{4}S + 2\right)$.

(12-19-20-17)

The knowledge that something remains yet un-enjoyed impairs our
enjoyment of the good before us.

Samuel Johnson

1. Two circles of different radii are tangent to each other. One of their common tangent lines intercepts them at two distinct points, A and B. If AB = $8\sqrt{2}$, and the smaller radius is 4, find the distance between the centers of the circles.

2. In problem #1, if R is the ratio between the area of the larger circle to the area of the smaller circle, find (5R).

3. Find the area, S for x > 0, bounded by y = - $\sqrt{x}$, y = - x², and the line x = 2. Then evaluate the expression: $\frac{30S(5+2\sqrt{2})}{17}$.

4. Find the volume, V, of the solid obtained by rotating the area between x + y = 2 and y = $\sqrt{x}$, for 0 ≤ x ≤ 1, around the y-axis. Then, evaluate: $[\frac{225V}{8\pi} + 2]$.

(12-20-20-17)

The happiest is he who suffers the least pain; the most miserable, he who enjoys the least pleasure.

Jean-Jacques Rousseau

1. A circle of radius $6(\sqrt{3} - 1)$ is inscribed in a 30°, 60°, 90° right triangle. Find the shortest side of the triangle.

2. If h is the hypotenuse of the triangle in problem #1, evaluate $(h - 3)$.

3. Find the volume, V, of the solid formed by rotating $y = 4 - x^2$ around the x-axis for $0 \le x \le 1$. Then find the exact value of:
$$\frac{300V}{203\pi}$$

4. Evaluate the integral $I = \int_0^1 \frac{(\arctan x)^{\frac{1}{2}}}{(x^2 + 1)} dx$. Then simplify: $\frac{180l}{\pi^{\frac{3}{2}}}$.

(12-21-20-15)

Do not spoil what you have by desiring what you have not; remember that what you now have was once among the things only hoped for.

Epicurus

1. In triangle ABC, DE is parallel to BC, with D and E on AB and AC respectively. If DB = 10, AC = 33, and EC = 15, find AD.

2. In problem #1, find BC if AD = DE.

3. Evaluate the integral $I = \int_4^5 \frac{x+1}{(x^2 - x - 6)} dx$. Then simplify: $\frac{100I}{\ln\left(\frac{56}{3}\right)}$.

4. For the family of functions $f(x) = \int x^3 \cos x\, dx$, find the specific value of the constant of integration, C, such that the point (0, 9) is on the graph of the respective function.

(12-22-20-15)

*Take full account of the excellencies which you possess, and in grati-
tude remember how you would hanker after them, if you had them
not.*

Marcus Aurelius

1. The volume of a cube is 216. Find the diagonal of the cube, D,
   and then, evaluate: $\frac{2D\sqrt{3}}{3}$.

2. If d is the diagonal of a side of a cube of total area 216, evaluate
   $(2d\sqrt{2} - 1)$.

3. Use implicit differentiation to find the equation of the tangent
   line to the curve $x^2 + y^2 = 9$ for x = 1 and y ≥ 0. If b is the
   y-intercept of this tangent line, evaluate: $\frac{40b\sqrt{2}}{9}$.

4. In problem #3, if X is the x-intercept of the tangent line, find
   $\left(\frac{5}{3}X + 2\right)$.

(12-23-20-17)

Think of all the beauty still left around you and be happy.

Anne Frank

1. The volume of a sphere inscribed in a cube is 288π. Find the
   side of the cube.

2. The volume of a sphere circumscribed to a cube is $864\pi\sqrt{3}$.
   Evaluate (2S), where S is the side of the cube.

3. Use implicit differentiation to find the equation of the tangent
   line to the curve $x^2 - 4x = 4y - y^2 - 4$ at the point
   $(2 + \sqrt{2}, 2 + \sqrt{2})$. If Y is the y-intercept of the tangent line, find:
   $5Y(2 - \sqrt{2})$.

4. In problem #3, if X is the x-intercept of the tangent line,
   evaluate: $\frac{15(2-\sqrt{2})}{4}$ X.

(12-24-20-15)

Nobody has things just as he would like them. The thing to do is to make a success with what material I have. It is a sheer waste of time and soul-power to imagine what I would do if things were different. They are not different.

Dr. Frank Crane

1. In a right rectangular prism the height is equal to the shorter side of the base, and it is half of the diagonal of the largest side of the prism. If the height is $\frac{12\sqrt{5}}{5}$, find the diagonal of the prism.

2. If R is the radius of the sphere circumscribed to the prism from problem #1, evaluate (4R + 1).

3. If A is the area bounded by $x^2 - y + 2 = 0$ and $y - x = 0$ for $x \in [0, 1]$,   evaluate $\frac{120}{11}A$.

4. Find V, the volume of the solid formed by rotating the area from problem #3 around the x-axis, and then, evaluate: $[\frac{75V}{26\pi} + 2]$.

(12-25-20-17)

Many are stubborn in pursuit of the path they have chosen, few in pursuit of the goal.

Friedrich Nietzsche

1. The volume of a right pyramid with an equilateral triangular base is 144cm³. If the height of the pyramid is $4\sqrt{3}$cm, find the side of the base.

2. If R is the radius of the circle circumscribed to an equilateral triangle of side 12, evaluate: $2(1 + R\sqrt{3})$.

3. A 6ft-tall person is walking away from a street light which is 24ft above the ground. If the person's constant speed is 6ft/sec, find the speed, S, of the tip of the person's shadow as it is moving away from the light post. Then, evaluate 2.5S.

4. In problem #3, find the rate of increase, r, of the distance between the head of the person and the tip of the person's shadow when the person is 24ft away from the light post. Then, evaluate $[\frac{75}{8}r + 2]$.

(12-26-20-17)

Where no plan is laid, where the disposal of time is surrendered merely to the chance of incident, chaos will soon reign.

Victor Hugo

1. The volume of a right pyramid with a square base is 256in^3. If the diagonal of the base is $8\sqrt{2}$in, find the height of the pyramid.
2. The volume of a right cone of height 8 is 96π. Find $\frac{9}{4}$ of the diameter of the base.
3. Two straight streets are perpendicular to each other. From their intersection, two cars depart one on each street. If the cars' constant speeds are 30mph and 40mph respectively, find the rate, r, at which the distance between the cars increases at the two-hour mark after departure. Then, evaluate (0.4r).
4. Find the rate r in problem #3, at the three-hour mark. Then, evaluate (.3r + 2).

(12-27-20-17)

There is in the worst of fortune the best of chances for a happy change.

Euripides

1. A right cone is inscribed in a sphere such that the base of the cone is the largest circular section in the sphere. If the volume of the cone is 72π, find the diameter of the sphere.
2. In problem #1, if r is the ratio of the volume of the sphere to the volume of the cone, calculate (7r).
3. A water container has the shape of a base-up right cone with the radius of the base 6in and the height 12in. If water is poured in at a rate of 4in^3/sec, find the rate, r, at which the water level rises when the depth of the water is 8in. Then, find $(80\pi r)$.
4. A car starts from rest at a constant acceleration of 2ft/sec^2. Find the distance, d, traveled by the car in 20 seconds. Then, evaluate: $[\frac{3d}{80} + 2]$.

(12-28-20-17)

The world is all gates, all opportunities, strings of tension waiting to be struck.

Ralph Waldo Emerson

1. The area of a side of a cube is 192. Find the radius of the sphere circumscribed to the cube.

2. If d is the diagonal of the cube in problem #1, evaluate (d + 5).

3. For the function $f(x) = \frac{x^2}{x-1}$, find its local maximum and minimum, and find the precise area, A, between the oblique asymptote of f(x) and the graph of the function for $x \in [2, 5]$. Then, evaluate: $[\frac{10A}{\ln 2}]$.

4. If $g(x) = \frac{(x^3-1)}{x+1}$, use limits, asymptotes, and derivatives of g(x) to sketch a graph of the function. Find the exact area, A, between the quadratic asymptote of g(x) and the graph of the function for $x \in [0, 1]$. Then, evaluate: $[\frac{7.5A}{\ln 2} + 2]$.

(12-29-20-17)

Who dares nothing, need hope for nothing.

J.C.F. von Schiller

1. A square of side L is rotated 360° around its diagonal. If the volume of the solid formed this way is $288\pi\sqrt{2}$, find L.

2. An equilateral triangle is rotated 360° around its height to form a cone. If the volume of the sphere inscribed in the cone is $500\pi\sqrt{3}$, find the side of the equilateral triangle.

3. A right rectangular pyramid with the sides of the base 6 and 8 is rotated 360° around its height. Use integration to set up a way to find the volume of the resulting solid, and then, if this volume is $V = \frac{500\pi}{3}$, find the height of the pyramid.

4. Illustrate how integration can be used to find the lateral area, L, of the solid formed in problem #3. Then, find L by any method, and evaluate: $[\frac{3L\sqrt{17}}{85\pi} + 2]$.

(12-30-20-17)

Be of good cheer. Do not think of today's failures, but of the success that may come tomorrow. You have set yourselves a difficult task, but you will succeed if you persevere; and you will find a joy in over-coming obstacles. Remember, no effort that we make to attain something beautiful is ever lost.

Helen Keller

1. Graph the line $4x + 3y = 24$ in the first quadrant, and rotate the graph around the y-axis to form a cone. Find the diameter of the base of the cone.

2. The diagonal of the base of a right rectangular prism is 20, and the length of the prism is 4 longer than its width. If the volume of the prism is 1728, and the sum of the length, width, and height of the prism is S, evaluate $(S - 6)$.

3. Find the area, A, bounded by the graphs of
    $y = x^2 + 2$, $y = x + 1$, $x = 0$, and $x = 2$. Then, evaluate: $\frac{15A}{2}$.

4. Find the volume, V, of the solid formed by rotating the area in problem #3 around the x-axis. Then, evaluate: $[\frac{75V}{82\pi} + 2]$.

(12-31-20-17)

About the Author

Irie Glajar was born in Communist Romania in 1955, has graduated from the Babes-Bolyai University of Cluj-Napoca, Romania with a Master's degree in Mathematics and Computer Science in 1979, and defected from the dictatorial regime in 1981. After several months in an Italian political refugee camp, he immigrated to the United States of America and since September 1982 has been teaching undergraduate mathematics at both high school and college levels in Austin, Texas. In 2010 Mr. Glajar earned an additional Master's degree in Mathematics Education from The University of North, Baia Mare, Romania.

Over the years, he has participated and presented at many professional conferences in the U.S., Canada, and Romania, and published several educational articles both in the U.S. and Romania. After the 2007 publication of his first book "WE ARE ALL ONE, The End of All Worries: Scientific and Spiritual Testimonies to the Unity of All Things," the author published "TEACH FOR LIFE, Essays on Modern Education for Teachers, Students, and Parents." He continued with "ESCAPE TO FREEDOM, Chronicles of a Life on Two Continents, My Escape from Communist Romania, An Autobiography," followed by the publication of his "2015 INSPIRATIONAL MATHEMATICS CALENDAR AND DAY PLANNER" which is a collection of 1,460 undergraduate math problems from basic to complex, spread as four problems per day with their answers in the respective date. And, under the title "EDUCATION IN A CHANGING WORLD, Essays for a Better Life," Irie's 2016 book is a collection of 34 essays on teaching and learning.

Besides treasuring his family life and his teaching career at Austin Community College, Austin, Texas, Irie Glajar finds much satisfaction in hobbies such as music, gardening, pets, and sports. Along with sciences and philosophy, he is also deeply interested in metaphysics, religion, spirituality, and international travel, which provide constant inspiration for his teaching of mathematics and his vision of modern education in general.

Irie Glajar can be contacted via email at: ir_gl@yahoo.com.

www.ingramcontent.com/pod-product-compliance
Lightning Source LLC
Chambersburg PA
CBHW060021210326
41520CB00009B/954